# MEETING THE
# COMMUNIST THREAT

# Meeting the Communist Threat

## TRUMAN TO REAGAN

## Thomas G. Paterson

New York    Oxford
OXFORD UNIVERSITY PRESS
1988

## Oxford University Press

Oxford   New York   Toronto
Delhi   Bombay   Calcutta   Madras   Karachi
Petaling Jaya   Singapore   Hong Kong   Tokyo
Nairobi   Dar es Salaam   Cape Town
Melbourne   Auckland

and associated companies in
Beirut   Berlin   Ibadan   Nicosia

## Copyright © 1988 by Thomas G. Paterson

Published by Oxford University Press, Inc.,
200 Madison Avenue, New York, New York 10016

Oxford is a registered trademark of Oxford University Press

Library of Congress Cataloging-in-Publication Data
Paterson, Thomas G, 1941–
Meeting the communist threat.
Includes index.
1. United States—Foreign relations—1945–
2. Anti-Communist movements—United States—History.
3. World politics—1945–   .   I. Title.
E744.P3118   1988   327.73   87-20433
ISBN 0-19-504533-5

2 4 6 8 9 7 5 3 1

Printed in the United States of America
on acid-free paper

*For my sister,*
*Shirley Paterson Gilmore*

# Preface

The nation which indulges toward another an habitual hatred or an habitual fondness is in some degree a slave.

*President George Washington, 1796*

Nobody in the military system ever described them [Vietnamese enemy] as anything other than Communism. They didn't give it a race, they didn't give it a sex, they didn't give it an age. They never let me believe it was just a philosophy in a man's mind.

*First Lieutenant William L. Calley, Jr., 1971*

When it comes to anti-communism . . . the United States is highly susceptible, rather like a drug addict, and the world is full of ideological 'pushers.'

*Senator J. William Fulbright, 1971*

A Risky U.S. Equation: E + O = T (WM)—Exaggeration + Oversimplification = Trouble (With Moscow).

*George F. Kennan, 1981*

What the U.S. does best is to understand itself. What it does worst is understand others.

*Carlos Fuentes, 1986*

"Why are Americans so anti-Communist?" I heard this question often during my 1987 lecture tour of the People's Republic of China. The simplest answer, which the Chinese probably expected at a time when President Ronald Reagan was still championing his conservative revolution, could have been that a host of right-wingers—ideologues, religious fundamentalists, the moral majority, business magnates, the military brass, and "anti" groups (opposed to affirmative action, abortion, busing for racial desegregation, and federal environmental regulations)—had gained power and were now using the instruments of government to foist their extremism on the American people. But that answer begged the question, for it did not explain why Americans had become so receptive to the right-wing message with its staunch anti-Communist rhetoric. In short, the message itself, not just its advocates, is important particularly because it has been transmitted in such an exaggerated form.

Americans have been anti-Communist because they perceive Communism as truly threatening and alien. It is obvious that Communism contrasts sharply with American ideology and experience. Americans celebrate private ownership of property, private profit, individual initiative, and the free marketplace of capitalism. They value their bill of rights, especially their First Amendment freedoms of speech and religion, and their political system of representative democracy and constitutionalism. The United States record has been blemished, of course, as Communists frequently remind sometimes embarrassed Americans, for practice has often violated ideal. Still, Americans proudly, even arrogantly, proclaim their economic and political liberties and invite people of other nations to match them. Americans believe that Communism denies these liberties and subordinates people to a suffocating, state-dictated regimen that stifles initiative, stunts economic growth, and perpetuates the power of self-serving, dictatorial rulers. Americans have also been anti-Communist because they recoil from the Communist call for world revolution and protest Soviet support of revolutions in the Third World. This is quite understandable, because Americans have seen most revolutions as challenges to their vaunted position as "the haves." Finally, American anti-Communism has been intensified by the dismal history of Communist governments and the cruelties they have visited upon their people—witness Josef Stalin's purges, Fidel Castro's political repression, or Mao Zedong's Cultural Revolution.

Americans did not invent a Communist or Soviet threat. With the Soviets' ability to punish the United States through nuclear attack and to diminish United States influence through an increasingly global reach, Moscow and its Communist allies have in fact posed a threat to American interests and principles. But why have Americans exaggerated the Communist threat, imagining a Communist monster of overwhelming capabilities and unlimited ambitions that hungers for troublemaking and intends them harm? Americans have made the Communist adversary into something it has never been, claiming for it a strength that it has never possessed, finding it menacingly at work where it has never existed, blaming it for troubles it has never started, depicting it as a monolith it has never resembled, and attributing to it accomplishments it has never achieved.

Threats to the United States' dominant international position have been and remain legion: economic threats from trade wars and embargoes; strategic threats from the nuclear arms race and small-nation wars; political threats from the volatile Third World; environmental threats from deforestation, acid rain, and short-ages of clean water; unpredictable but relentless threats from terrorists; and threats from American clients and friends whose political instability, financial indebtedness, nationalistic aspira-tions, and independent decisions frequently make them un-steady allies. Most of these threats, in fact, have little to do with Communism or the Soviet Union. Nor have many of the world's disturbances had Communist sources. Tensions in the Middle East since the 1940s have stemmed not from Soviet or Communist intrigue but from profound Arab-Israeli competition, intra-Arab feuding, and bitter religious quarrels among Muslims. Troubles in Asia and Africa have derived not so much from Communist plotting as from rebellions aimed at breaking colonial bonds and from internal power struggles after independence. Disorder in Latin America has sprung not from Communist meddling but from deep-seated economic inequalities, class resentments, and the political dominance of landholding and entrepreneurial elites and military strongmen. And, despite the Tito-Stalin schism, the Sino-Soviet split, the limited but emerging independence of some Eastern European nations, Cuban-Soviet differences, and the demonstrated weakness of the Soviet appeal in the Third World,

Americans and their leaders continue to speak of "the Communist threat" as if "Communism" constituted a unified movement. They continue to see the Soviets and other Communists lurking behind most of the world's trouble spots.

Leaders often see or hear what they want to; they respond not only to what is real, but also to what they already believe based upon experience, inherited suspicions, and personal prejudices. When discrepant information about an "enemy" calls into question strongly-held beliefs, they strive to preserve those beliefs by discounting or distorting that information, refusing to think about it, engaging in wishful thinking, discrediting the source, or searching for other evidence confirming their beliefs. Americans and their leaders have also nurtured a vigorous anti-Communism because they have been notably ignorant and indifferent about world affairs. Such ignorance elevates "gut feelings" and intuition as substitutes for close analysis and spawns stereotyping and paranoia. In a chaotic, discomforting world, Americans have been prone to point an accusing finger at one conspiratorial enemy. Simplicity, after all, can provide clarity that sometimes misleads. Arrogance accompanies ignorance: Americans have traditionally seen themselves as the conveyors of material success and as noble missionaries in a world of sinners. In contrast, the "enemy" is painted in the blackest terms and the picture that emerges is simple: good versus evil.

Some leaders hold on to outmoded, wrongheaded views of the Communist threat because it is politically expedient to do so, or because they fear or respect interest groups that hoist the anti-Communist banner. Perhaps an organizational imperative drives some leaders: The budget-conscious American military establishment, for example, has earned notoriety from administration to administration by hyping the Communist threat. Some industrialists may contribute to hyperbolic renderings of the Communist threat because the resultant, large federal defense spending is good for business. Still other Americans, confused by conflicting governmental reports and scholarly disagreements, may conclude that we can never know the nature of the Communist threat and must therefore assume the worst. For many Americans, especially conservatives, exaggeration of the Communist threat affords the opportunity to reassert traditional American values and to reassure themselves of core ideas—to declare Americanism in the strongest possible terms by pillorying Communism in the strongest possible terms. Anti-Communism and threat exaggeration, in this sense, serve as vehicles for forging American unity at home, particularly in times of domestic crisis. Given these characteristics, leaders who have developed views of a malevolent Communism that preys upon a vulnerable world may not shift their views, even in the face of abundant evidence that Communism is not the omnipresent force they imagine. Because leaders work to maintain consistency in their ideas, they will often ignore contradictions, cling to exaggerations, and become intransigent.

In this book I have attempted to find out why and how Americans and their leaders have perceived, exaggerated, and devised policies to meet the Communist threat. I have explored why and how the United States has undertaken, from the 1940s to the present, global interventionism, the containment doctrine, nuclear arms expansion, covert activities, "model" economic development programs, and extensive foreign aid. By studying key Cold War assumptions and policies, critics of the prevailing wisdom, prominent personalities from Harry S Truman to Ronald Reagan, and crises in different regions since the 1940s, I have tried to demonstrate the tenacity, momentum, and permanence of a way of thinking about the world and the American place in it.

Exaggeration of the Communist threat has led to consequences Americans did not intend, and to activities that have violated their professed ideals of self-determination, democracy, and opportunity. They have assigned themselves the roles of world policeman, teacher, banker, and social worker; they have over-spent tremendously the federal budget and have diverted funds from domestic programs to pay for a huge military arsenal; they have built up massive quantities of nuclear-tipped weapons and have talked carelessly about using them; they have repeatedly dispatched troops to other nations; they have extended their economic power to the point of domination; and they have engaged in covert operations to undermine sovereign governments and assassinate their officials. In the post-1945 period, the over-estimation of the Communist threat has led to global containment, and, in turn, to American expansion and an empire that faces unrelenting challenges. Paradoxically, American global activism to extend and protect American interests and to guard against the spread of Communism has not produced more security, but rather a deeper vulnerability. The exaggeration of the Communist threat, in the end, has meant more danger and more threat.

An historian is so dependent upon others to help gather source material, write the story, and analyze its meaning. For reading parts of the manuscript, I thank Kenneth J. Hagan, Walter LaFeber, Elizabeth Mahan, Robert McMahon, and Nancy Tucker. At the University of Connecticut I have benefited from the research assistance of Barney J. Rickman III, Ann Balcolm, and Laura Grant. My colleague and friend J. Garry Clifford remains my best critic; *he* knows how to research, analyze, and write. The archivists at the presidential libraries and other depositories cited in the notes proved invaluable; I simply could not have found my way through the millions of documents without their professional guidance. Sondra Astor Stave prepared the index with noted competence. For financial assistance which has permitted me to reach distant archives and to take leaves from teaching and administrative responsibilities, I heartily thank the University of Connecticut Research Foundation, National Endowment for the Humanities, Institute for the Study of World Politics, American Philosophical Society, Harry S Truman Institute, Hoover Presidential Library Association, Gerald R. Ford Foundation, and Lyndon Baines Johnson Foundation. Some of the chapters in this book appeared earlier as articles. They have been revised to incorporate recent scholarship and to eliminate duplication, and I thank the journals and publishers for granting me permission to use previously printed material. To keep notes to a minimum I have included them only to indicate sources of quotations. Finally, I am grateful to my editors at Oxford University Press, Sheldon Meyer and Nancy Evans.

*Storrs, Connecticut, and*                                              T.G.P
*Monhegan Island, Maine*
*July 1987*

# Contents

# MEETING THE
# COMMUNIST THREAT

# 1

# Red Fascism: The American Image
# of Aggressive Totalitarianism

In the nervous early days of the Cold War, Herbert L. Matthews of the *New York Times* asked: "Should we now place Stalinist Russia in the same category as Hitlerite Germany? Should we say that she is Fascist?"[1] Yes, he answered—as did many other Americans in the post-World War II era. President Harry S Truman himself declared "there isn't any difference in totalitarian states. I don't care what you call them, Nazi, Communist or Fascist. . . ."[2] Both before and after the Second World War, Americans casually and regularly sketched similarities between Nazi and Communist ideologies, between German and Soviet foreign policies, authoritarian controls, and trade practices, and between Adolf Hitler and Josef Stalin. After the war, Americans easily transferred their hatred for Hitler's Germany to Stalin's Russia. "It is really a matter of labels," wrote Matthews.[3] The label that gained currency was "Red Fascism."[4] Popularizers of the phrase seemed to be seeking relief from their frustrated hopes for a peaceful postwar world and from their dismay in finding growing international tension after a long and destructive conflict. The analogy depicting the two European nations as twin totalitarian states seemed to assure perplexed and alarmed Americans that they knew what to expect from postwar Russia, because the analogy told them that the 1940s and 1950s would simply be a replay of the 1930s. They were not.

Americans had long used expressions like despotism, Prussianism, dictatorship, autocracy, tyranny, and authoritarianism to describe anti-democratic governments that were anathema to the American way. These words, like "totalitarianism," conjured up images of absolute state power, suppression of civil liberties, denial of religious freedom, submersion of the individual to the purposes of the state, reliance on force and militaristic aggression, imperialism, systematic terror, lies and deceit, and self-conscious ideology. Totalitarianism was the antithesis of Americans' self-image and principles, but their own national experience left them unprepared to give the word careful definition. Until the Italian attack on Ethiopia and the rise of Hitler "gave Fascism a demonic image," some Americans actually flirted with Fascist Italy as an attractive social and political experiment.[5] The term "totalitarianism" itself was imported from Europe and first applied to Mussolini's Italy and Hitler's Germany. Though coined in the 1920s, the word did not come into general or academic use until the late 1930s. Communist leaders preferred to call their system "socialist democracy." But in the late 1930s some anti-Communist writers began to apply "totalitarianism" to Soviet Russia and its one-party government. During World War II the word served the purposes of anti-Nazi propaganda; after the war it became an "anti-Communist slogan."[6]

As a tool of propaganda, the term "totalitarianism" lost its utility for systematic analysis of different political systems and their methods. While recognizing this problem, the diplomat and historian George F. Kennan has argued persuasively that for totalitarianism "there are at least no better examples than Germany and Russia."[7] It is nevertheless true that, because the outward appearances of the two systems seemed to be more similar to each other than to any previous political system in the world, differences between Fascist and Communist systems and their behaviors were obscured. Ignoring the widely diverse origins, ideologies, goals, and practices of totalitarian regimes, Americans tended to focus on the similar methods employed by such regimes and to assume that these methods, practiced in the

1930s by Hitler, would be repeated necessarily in the 1940s by Stalin. This oft-heard analogy, sometimes factually based but often distorted and misleading, carried strong moral appeal and influenced American policymakers toward an uncompromising Cold War posture.

It is understandable that Americans would be drawn to the simple analogy in the face of the Soviet Union's imperialistic behavior after the Second World War in Eastern Europe, but it is also noteworthy that such thinking was more the stuff of propaganda, exaggeration, and international gamesmanship than of patient inquiry and statesmanship. Totalitarian states have, of course, exhibited undeniable similarities. Yet it did not follow that Stalin's Russia would or could always act like Hitler's Germany or that Russia was set inexorably on a path of military aggression. But what is more important here is not that Nazi Germany or Communist Russia were different or similar, but that many Americans took the unhistorical and illogical view that Russia in the 1940s would behave as Germany had in the previous decade because of the supposedly immutable characteristics of totalitarians. It is always difficult to demonstrate the power of an idea or image, but this chapter establishes that the "Red-Fascist" idiom was widely articulated and believed. To understand the American reading of foreign events and United States policies in the early Cold War years we must explain the pervasive presence of the analogy.

Through the 1930s, until the Nazi-Soviet Pact of 1939, Americans who made comparisons gave a mixed picture. Some stressed differences in ideology and diplomatic posture; whereas Hitler's Germany was openly hostile to Western democracies, Soviet Russia sought a united front against Fascism. But a general prewar consensus emerged that highlighted similarities, especially in the methods the two states used to perpetuate their power. The historian Charles Beard criticized the elitism he perceived in both Fascist and Communist dictatorships; Archibald MacLeish condemned both systems for stifling intellectual freedom; and Ambassador to Germany William E. Dodd concluded that both Hitler and Stalin practiced militarism to keep unemployment down. After the purge trials in Russia and the persecution of the Jews in Germany, Elmer

Davis, John Dewey, Walter Lippmann, and editors of the *New York Times* spoke out against the undemocratic, totalitarian similarities in Germany and Russia. The differences between the two, growled the *American Mercury* magazine, were like the differences between the poisons arsenic and strychnine.

Though American observers had frequently commented on the lack of a representative government in Russia after 1917, Soviet secretiveness, censorship, disregard for public opinion, purges, ideological purification, and frenzied denunciation of enemies in the 1930s seemed to parallel characteristics of the Nazi regime. The conservative journalist Eugene Lyons, disillusioned after his failure to find a Soviet "utopia" during his stay in Russia, took his readers on a tour of European tyrannies—"totalitarian insanities"—which he equated with Russia. Everywhere he saw "the autocrats using almost the identical slogans, wielding the selfsame 'sword of history' for class or race or nation." He lamented the "moral collapse of Europe," the decline of humanistic values asserting the dignity of life and a respect for truth. And he asked: "What is to distinguish Socialism according to Stalin from Socialism according to Mussolini?"[8]

Like Lyons, many Americans blurred the ideological differences between Communism and Fascism and tended to believe that totalitarian methods overrode ideology in shaping political forms. The scholar Hans Kohn wrote forcefully in his *Revolutions and Dictatorships* (1939) against this distortion of ideology, but scholarly opinion, like public opinion, increasingly moved in the opposite direction. More than a decade later Hannah Arendt argued the majority opinion in her influential book, *The Origins of Totalitarianism*. She saw a "complete indifference to mass interest" as the guiding characteristic of the German and Russian totalitarian regimes.[9] Yet she made no distinction, critics have noted, between one system proclaiming a humanistic ideology and failing to live up to its ideal and the other system living up to its anti-humanistic and destructive ideology only too well, and between the modernization of an advanced industrial nation (Germany) and an essentially underdeveloped nation (Russia).

With the profoundly disturbing news in late August 1939 that the German Reich and the Soviet Union had signed a mutual non-aggression pact, and with the subsequent German and Russian invasion and division of Poland, the most significant prewar identification of the two regimes was established. "Hitlerism is brown communism, Stalinism is red fascism," wrote the *New York Times*.[10] Some liberals, such as Vincent Sheehan and Louis Fischer, who had held out hope that Russia would avoid totalitarianism, now concluded that Hitler and Stalin were full-fledged partners, and, according to Sheehan, that Stalin had embraced Fascism. Religious opinion—Protestant, Catholic, and Jewish—also strongly denounced the Nazi-Soviet Pact. The Reverend A. J. Muste, a liberal pacifist, concluded that the two states were anti-capitalistic, anti-Christian, and anti-democratic and foresaw a "vast historical movement" toward their merger.[11] American Jews feared that Stalin would initiate Hitler-like policies in Poland and recalled that Russian anti-Semitism had deeper roots than that of the Germans. One Jewish journal, *The Reconstructionist*, referred to the anti-religious efforts of both Russia's Militant Atheist League and Germany's Gestapo as twin attempts at "spiritual liquidation."[12] Russia's unprovoked attack on Finland in 1939 also aroused American indignation; it was now clear that both Germany and Russia were aggressors in Europe. War relief crusades for the Finns produced an ecstatic American response. Robert Sherwood responded with his drama, *There Shall Be No Night*, condemning the German and Soviet aggressive conspiracy against world democracy. Frederick Hazlitt Brennan in February 1940 popularized the phrase "Commu-Nazi" in a five-part *Collier's* story called "Let Me Call You Comrade."

Thus, on the eve of the Second World War, many Americans linked Fascist and Communist ideologies as denials of human freedom, saw Germany and Russia as international aggressors, and pictured Hitler and Stalin as evil comrades. Shortly after the sudden German invasion of Russia in June 1941, the *Wall Street Journal* indicated its ambivalent position on the outcome of the new war: "The American people know that the principal difference between Mr. Hitler and Mr. Stalin is the size of their respective mustaches."[13] Former Ambassador to Russia William C. Bullitt saw the new contest as one between "Satan and Lucifer," and Senator Harry S Truman of Missouri made one of his famous utterances: "If we see Germany winning we ought to help Russia and if Russia is winning we ought to help Germany and that way let them kill as many as possible, although I don't want to see Hitler victorious under any circumstances. Neither of them think anything of their pledged word."[14] Some American isolationists also denounced the power politics of both Germany and Russia and adopted a plague-on-both-your-houses attitude. Yet, after the German invasion, President Franklin D. Roosevelt, against ardent opposition, extended Lend-Lease aid to Russia. The opinion of most interventionists was that, though Russia was evil, it was at least now fighting evil.

After the entry of the United States into World War II, Americans focused on the differences between Hitler's Germany and Stalin's Russia in order to help cement the wartime alliance. It now became popular to stress that Russia and the United States were similar; both were anti-imperialist and both had experienced a revolutionary past. *Collier's*, superficial as usual, actually concluded that Russia was "evolving from a sort of Fascism . . . toward something resembling our own and Great Britain's democracy."[15] But the emphasis on differences proved a temporary façade, a reaction to Soviet war efforts rather than a reappraisal, and the Nazi-Communist analogy appeared again as Soviet-American tensions increased near the close of the war.

Even before the war ended Ambassador W. Averell Harriman suggested that the thrust of Communism was not dead and that the United States might have to confront an ideological war perhaps as "vigorous and dangerous as Fascism or Nazism."[16] Acting Secretary of State Joseph C. Grew advised President Truman in June 1945 that "Communists have the same attitude as Goebbels did—that the civil liberties of democracies are convenient instruments for Communists to facilitate their tearing down the structure of the state and thereafter abolishing all civil rights."[17] To those who ridiculed his call for a study of Soviet philosophy, Secretary of the Navy James Forrestal replied that "we always should remember that we also laughed at Hitler."[18] A New Hampshire lawyer, later to be that state's Attorney General and congressman, warned against Moscow's revolutionary goals and asked in 1946: "Do you remember that

Hitler's plans were fully outlined in his book *Mein Kampf* and that no one paid serious attention?"[19] To Americans, Stalin became a new Hitler—demagogic, dictatorial, demanding personal loyalty, conniving to rule other peoples. The tough but friendly "Uncle Joe" of wartime propaganda became the paranoid tyrant of the Cold War, aping Hitler.

In totalitarian states, Americans were aware, absolute control over the means of communication meant that the regime gave people only the information it wished them to have. Germany had controlled information, and a grim Grew lectured a nationwide radio audience in June 1945 that "never again must a tyranny be permitted to mislead and befuddle a people and to betray men and women into mob violence, aggression, and national suicide."[20] Many Americans in the postwar period believed that Russia's control of communications, information, and propaganda merely replicated Joseph P. Goebbel's German model, and many assumed that aggressive war would flow from an absence of free expression. General John R. Deane, the head of the American military mission in wartime Russia, assumed mental and physical regimentation as a characteristic of totalitarianism. His much-publicized *Strange Alliance* (1947) drew heavily upon the Nazi-Soviet analogy. He noted that the marching of Russian soldiers "closely resembled the goose-step, with arms rigid and legs kicked stiffly to the front," and this "pointed plainly to a discipline oriented toward German methods, which tends to destroy individual initiative in the battle pay off." Deane also related this regimentation to the control of ideas. "Unfortunately the Russian people are not allowed to see that the pattern being cut by their leaders is much the same as that which was followed in Germany."[21] President Lewis H. Brown of the Johns-Manville Corporation agreed that the "Russian people, like the German people, do not want to rule the world, but they are helpless slaves of the ruling clique that dominates the people through fear and terror, through concentration camps and secret police and through the whole mechanism of totalitarianism."[22]

Control through fear and terror became, indeed, a significant component of the totalitarian image. Americans knew of Russian exile and labor camps in Siberia even before the Bolshevik

Revolution in 1917, and in the 1920s and 1930s it was known that such camps were filled with political prisoners, criminals, and those opposed to Soviet collectivization schemes. The German experience, however, seems to have stamped the image of the concentration camp, with all its overtones of mass extermination and unbridled horror, on the Russian camps. Congresswoman Clare Boothe Luce castigated the Soviet system as one "which keeps eighteen million people out of 180 million in concentration and forced labor camps."[23] In the United Nations, the American delegate Willard L. Thorp compared the "shocking exploitation of human beings by the Nazis" with forced labor conditions in Russia.[24] In 1947 Senator J. Howard McGrath, later Truman's Attorney General, applied the analogy of concentration camps to Eastern Europe and, citing as his source the vehemently anti-Communist publication *Plain Talk* and the writings of anti-Soviet writers such as William C. Bullitt, McGrath claimed that the clergy in Estonia, Latvia, Lithuania, and Yugoslavia were being exterminated. The image of Nazi death camps was thus created; so was that of the police state. Arthur Bliss Lane, American Ambassador to Poland from 1945 to 1947, lectured that the Russian security police copied Gestapo tactics. Speaking of persons brutally beaten and tortured by police, Lane told a radio audience that "the same terror of a knock at the door in the dead of night exists today as it did during the Nazi occupation."[25]

Americans pointed out too that both Fascist and Communist regimes attempted to extend their ideological appeal and brute tactics to other nations through subversive agents. Sensational spy cases in Canada and the United States inflated fears of foreign agents. Americans credited Soviet Communists with better "fifth-column" activities than those of the Nazis. In June 1945 the State Department informed the President that "a communist party was in fact a fifth column as much as any Bund group, except that the latter were crude and ineffective in comparison with Communists."[26] Swayed by this analogy, for example, American leaders could only perceive the civil war in Greece as a Hitler-like fifth-column intrusion by the Russians and not, as it was in reality, a struggle of Greeks against a British-supported monarchy with negligible interference by the Soviet Union.

Perhaps the most significant, and the most misleading, part of the Nazi-Communist analogy was that drawn between the prewar and wartime military actions of Germany and those of Russia in the postwar period. As Soviet armies stormed through Eastern Europe on the heels of the defeated *Wehrmacht*, many Americans interpreted the victories as immediate aggression rather than as wartime liberation. Americans assumed that Russia was simply replacing Germany as the disrupter of peace in Europe. They did not understand intense Soviet security concerns aroused by the recent German ravaging of the Russian countryside following the march through Polish territory, which left millions dead. The term "satellite," first applied to German domination of Rumania and Hungary, was easily transferred to Russian hegemony in postwar Eastern Europe. Winston Churchill, who helped popularize the notion in America, lumped Germany and Russia together as similar aggressors, and both Max Eastman and the Russian emigré Ely Culbertson condemned Russia for employing the German practice of disregarding treaties and adopting satellite states. H. V. Kaltenborn, shortly after Churchill's famous "Iron Curtain" speech in 1946, warned his radio listeners to "Remember Munich!"[27] The exaggerated view of many public opinion leaders was that Russia, like Germany before, was going to sweep over Europe in a massive military attack, and then reach to other continents. MacGrath told the Senate: "Today it is Trieste, Korea, and Manchuria, tomorrow it is the British Empire. The next day it is South America. And then—who is so blind as to fail to see the next step?"[28] His words clearly suggested the parallel between prewar aggression and postwar Russia and helped make the case for military containment through alliances like the North Atlantic Treaty Organization (NATO).

George F. Kennan, State Department expert on the Soviet Union, attempted in 1956 to dispel a myth that he himself had helped create years earlier (see Chapter 7). "The image of a Stalinist Russia," he argued, "poised and yearning to attack the West, and deterred only by our possession of atomic weapons, was largely a creation of the Western imagination."[29] Kennan has claimed that the containment doctrine he advocated in private and in his influential "X" article in the July 1947 issue of *Foreign Affairs* did not recommend military containment, the creation of a ring of bases and alliances around Russia, or the identification of German aggression with the Russian presence in Eastern Europe. Yet popularizers of the Nazi-Soviet analogy in official Washington and elsewhere used Kennan, in part because of his own imprecision in 1947, to argue their case that Russian "aggression" had to be halted or America would face another world war. Protesting what he considered to be the misuse of his ideas, Kennan could only conclude that "Washington's reactions" had been "deeply subjective."[30]

"Munich" and "appeasement" returned as terms of shameful "sell-out" to haunt postwar negotiations with the Soviet Union. Convinced that President Roosevelt had conceded too much at the Yalta Conference, Senator Arthur Vandenberg, in San Francisco at the United Nations meeting, told his diary that "there is a general disposition to *stop this Stalin appeasement. It has* to stop *sometime.* Every surrender makes it more difficult."[31] In defending the Truman Doctrine in 1947, he remarked that "I think the adventure is worth trying as an alternative to another 'Munich' and perhaps to another war. . . ."[32] To the suggestion made at a Cabinet meeting in September 1945, that the United States eliminate its monopoly of atomic bombs and nuclear information in the interests of peace, Forrestal replied, "It seems doubtful that we should endeavor to buy their understanding and sympathy. We tried that once with Hitler. There are no returns on appeasement."[33] *Barron's* chastised Secretary of Commerce Henry Wallace for having an "appeaser's dream" when he advocated atomic disarmament through negotiations with Russia.[34] General Douglas MacArthur upbraided those in the administration who would not escalate the Korean War, for they were adhering to "the concept of appeasement, the concept that when you use force, you can limit the force."[35] Because the cry of appeasement was pervasive in the American mind, diplomats may have been less willing to negotiate and bargain with the Soviet Union. Indeed, for some, diplomacy and appeasement were probably nearly identical in meaning, and diplomacy with totalitarian states necessarily meant concession of principle. President Truman suggested as much in his Navy Day speech of October 1945 when he stated that "we shall firmly adhere to what we believe to be right; and we shall not give our approval to any compromise with evil."[36] With echoes of "appeasement" and "Munich" heard in the august rooms of the White House, it is no wonder that the American President never directly negotiated with the chief Soviet leader from 1945 (Potsdam) to 1955 (Geneva).

The State Department's publication on January 21, 1948, of captured German documents related to the Nazi-Soviet Pact of 1939 fed the notion that Russia was aggressive, deceitful, and opportunistic—a clone of Nazi Germany. Walter Lippmann thought the publication "the work of propagandists and not of scholars," but most commentators read the selected documents as validation of the charge that the Nazis and the Russians had been essentially one and the same in their prewar aims of world conquest.[37] Dorothy Thompson waxed hyperbolic when she expressed bafflement that the American government had gone ahead with the United Nations Charter and Nuremberg Trials "with the Russians sitting as prosecutors and judges against the very persons they had egged on to war and with whom they had plotted to divide the spoils."[38] Ignoring the history of Russia's rebuffed efforts to form an anti-German coalition in the 1930s—an ignorance reinforced by the selected published documents—the *New York Times* editorialized on the basis of the German documents alone that "the initiative toward the conspiracy did not come from the Nazis, but from Moscow, behind the backs of France and Britain," and Bertram D. Hulen of the *Times* thought the documents proved that Soviet officials "would rather work with the Germans than with the West."[39] Kaltenborn believed the publication of the documents in early 1948 was a maneuver by the Truman Administration to push Congress into passing Marshall Plan legislation. Secretary Marshall rebutted, however, that the publication was routine and had in fact been postponed pending the results of the Foreign Ministers' Conference of December 1947 in order not to offend the Russians; with the failure of the meeting, the documents were released. More important than the question of the timing is the significance of these authentic historical materials to the development of American Cold War thought. The documents reinforced and reflected the American image of the Nazi-Soviet connection and strengthened the argument of those who believed that Russia had never shared Allied war goals, but rather embraced aims of world domination.

International trade was another element in the Nazi-Soviet comparison. Russia conducted foreign trade through the agency of the government, as had Germany before. Both, it was suggested, used trade for political purposes, and both imposed harsh commercial treaties on Eastern European countries. Germany and Russia thus forced weaker nations to buy goods at exorbitant prices and to sell products to them at reduced rates. Trade, then, was another weapon in the aggressive arsenal of totalitarian states. Recalling American trade with Germany and Japan before the war, Kennan stated in 1945 that there was little to gain, and much to lose, from postwar American-Russian trade. By trading with postwar Russia, he reasoned, Americans might be "furthering the military industrialization of the Soviet Union," and "be creating military strength which might some day be used to our disadvantage. . . ."[40] His conspicuous suggestion was that prewar American trade with the Axis Powers had served to build up the enemy against the United States, as it may have, and that postwar Russian-American trade might replay that mistake. Such thinking forced rigidity upon postwar American foreign policy, because it assumed that the course of relations was already set and that the prewar decade provided an accurate map for the postwar era.

In 1949, Professor Leo Szilard of the University of Chicago, one of the builders of the atomic bomb, wrote: "Soviet Russia is a dictatorship no less ruthless perhaps than was Hitler's dictatorship in Germany. Does it follow that Russia will act as Hitler's Germany acted?"[41] Szilard did not think so, and his question emphasizes at once the major assumption and the major weakness of the Nazi-Communist analogy: that deep-seated conflict with totalitarianism was inevitable after World War II; that there was no room for accommodation with the Soviet Union because the Communist nation was inexorably driven by its ideology and its totalitarianism. It followed from such reasoning that the United States could have done nothing to alleviate postwar tension. Such a notion ignored the years 1945–46 when the possibilities for accommodation were far greater than later in the decade. The "Red-Fascist" analogy itself obstructed accommodation: it did not allow for a sophisticated understanding of Soviet intentions or capabilities and power relationships in Europe, it substituted emotion for intellect, and it particularly distorted the American perception of reality, ballooning the Soviet threat. Thus, prepare for the worst, get tough, and avoid compromise. The analogy taught, in short, that the enigma of Soviet Russia could be fathomed only by the application of the historical lessons learned in the 1930s.

A few prominent American political figures in the early postwar era, such as Henry Wallace, consistently refuted the analogy. He asked how America could deal with the undesirable reality of Russian hegemony in Eastern Europe—a hegemony Wallace defined as defensive rather than aggressive. Wallace wrote to Truman in July 1946 that Russia had legitimate security needs in Europe. "We should be prepared," he asserted, "even at the expense of risking epithets of appeasement, to agree to reasonable Russian guarantees of security."[42] Other critics noted that Americans drew little distinction between the German drive for European domination and the Soviet interest in supporting indigenous revolution—between military attack and internal revolution. And, to mention another difference, Marxian philosophy, however misapplied, looked for social and economic improvement among disadvantaged people, whereas Fascism was designed not to improve mankind but rather to destroy that part it disliked. The American clouding of distinctions between military Fascism and revolutionary Marxism contributed to the establishment of world-wide alliances and major military containment policies in Europe and Asia. The Hitler-Stalin comparison was also superficial and misleading. Kennan himself later attempted to convince his readers that Stalin's intentions, though menacing in Western eyes, were "not to be confused with the reckless plans and military timetable of a Hitler."[43] Brutal as Stalin was, there is little evidence, as Kennan has indicated, to suggest that he was a madman bent on world conquest. There was no Soviet "blueprint" for expansion in the postwar period. Americans, enthralled with "Red Fascism," exaggerated the Soviet threat, blaming Moscow for the postwar world's ills and attributing to it expansive intentions and capabilities that Moscow's behavior never matched, reprehensible as that behavior often was.

The American image of "Red Fascism" embraced emotion and simplism, and the compelling fictional creations and anti-utopias of writers such as George Orwell, Aldous Huxley, and Arthur Koestler helped foster the superficial analogy. Orwell's *1984*, appearing at a time when American fears of totalitarianism had become obsessive, did much to shape American opinion. The image of totalitarianism presented in *1984* became a model, as unreal and probably as significant as that created by American leaders and the mass media from the war's end to the book's publication in America in 1949. So closely had the Nazi-Soviet image been woven into American thought that it proved difficult for many Americans to read the book without applying totalitarian stereotypes from the Nazi-Soviet analogy. A *Life* editorial, reprinted along with a condensed version of *1984* in *Reader's Digest*, found the book "so good, indeed, so full of excitement and horror, that there is some danger that its message will be ignored." Clarifying the novel's message, *Life's* editors unhesitatingly identified the central and alarming figure of "Big Brother" as a "mating" of Hitler and Stalin and made it clear that Russia and Germany were to be substituted for the author's use of London as the novel's setting, an interpretation perhaps more indicative of American perception than of Orwell's own intentions.[44]

It was in Kennan's introspective mind that the impact of the analogy and image was best understood and articulated. Well aware of the components of the analogy, many of which he believed with a majority of Americans, Kennan also recognized the additional dream-like quality that the Red-Fascist image had taken in the American mind:

> When I try to picture totalitarianism to myself as a general phenomenon, what comes into my mind most prominently is neither the Soviet picture nor the Nazi picture as I have known them in the flesh, but rather the fictional and symbolic images created by such people as Orwell or Kafka or Koestler or the early Soviet satirists. The purest expression of the phenomenon, in other words, seems to me to have been rendered not

in its physical reality but in its power as a dream, or a nightmare. Not that it lacks the physical reality, or that this reality is lacking in power; but it is precisely in the way it appears to people, in the impact it has on the subconscious, in the state of mind it creates in its victims, that totalitarianism reveals most deeply its meaning and nature. Here, then, we seem to have a phenomenon of which it can be said that it is both a reality and a bad dream, but that its deepest reality lies strangely enough in its manifestation as a dream. . . .[45]

This nightmare of "Red Fascism" bestirred a generation of Americans, leaving its mark on the events of the Cold War and its warriors.

# 2

## America's Quest for Peace and Prosperity: European Reconstruction and Anti-Communism

If Soviet Communism was destined, as the Red-Fascist notion imagined, to repeat the ugly cycle of the 1930s economic depression, German totalitarian aggression, and global war, some thought it would try to accomplish its malevolent purposes not by sending the vaunted Red Army into combat but rather by preying on postwar economic instability and ensuing political chaos. "World Communism is like a malignant parasite which feeds only on diseased tissue," Kennan told his State Department superiors.[1] Drawing upon the lessons of the depression decade, Americans vowed never to duplicate that wrenching past. The concept of Red Fascism prophesied a foreboding future; the concept of "peace and prosperity" explained how that future could be brightened. American leaders set a course of preventing or eradicating depression and war—and hence totalitarianism and Communism—through postwar reconstruction policies and a revival of world trade. Swelled by economic power when much of the rest of the world lay in ruins, the United States determined to use foreign aid as an instrument to meet the Communist threat and at the same time insure American expansion abroad, which Americans believed would benefit all.

The ideology of "peace and prosperity" enjoyed wide accep-

18

tance among segments of the political spectrum, although they differed over tactics. But, from the then liberal magazine *The New Republic* to the conservative National Association of Manufacturers, from Henry A. Wallace to President Truman, from labor union officers to corporate executives, Americans agreed that economic instability and poverty bred political instability, revolutionary politics, totalitarianism, violence, aggression, and finally war, and that only the United States had the power to break the terrible pattern. This thinking was simple and rigid, and it joined Red Fascism in exaggerating the Communist/Soviet threat and in contributing to a postwar expansion of American influence overseas that amounted to a self-fulfilling prophecy: the Soviets, already a nuisance, became more so in reacting to a United States expansion they considered "encirclement."

As American leaders defined it, the peace and prosperity theme had several interlocking components which directed foreign policy toward extensive foreign aid and vigorous foreign trade, both of which were realized in the Marshall Plan. The first premise read that world peace required global economic well-being. "Hungry people are not reasonable people," concluded a State Department official. "Their thoughts are concerned with their own misery and particularly with the tortured cries of their hungry children. They are easy victims of mass hysteria."[2] Indeed, destitute people were prone to accept totalitarian and undemocratic political movements. State Department Counselor Benjamin V. Cohen agreed that "people long tired, cold, hungry, and impoverished are not wont to examine critically the credentials of those promising food, shelter, and clothes."[3] The Economic Cooperation Administration found the postwar economic recovery of Western Germany necessary "to avoid the political dangers which might result from economic distress."[4]

Peace in the American definition meant political and social order, and as W. Averell Harriman, former Ambassador to the Soviet Union and Secretary of Commerce, put it, "political stability can only be attained through improving economic conditions."[5] The staff of the Council of Economic Advisers wrote similarly about conditions in Europe: "The lingering economic illness . . . retards the attainment of the political stability which is the prerequisite to the maintenance of free institutions." The council itself urged foreign aid in late 1947 for Italy and France, because industrial paralysis there might mean "serious social disturbance and political reorientation."[6] Under Secretary of State Will Clayton, after a visit to Europe in early 1947, wrote an influential report on the continent's needs: "Europe is steadily deteriorating. The political position reflects the economic. One political crisis after another merely denotes the existence of grave economic distress."[7] "It is logical," said Secretary Marshall in his famous speech at Harvard University on June 5, 1947, "that the United States should do whatever it is able to do to assist in the return of normal economic health in the world, without which there can be no political stability and no assured peace."[8] In short, economically-hobbled people tended to be radical in politics, capable of being exploited by Communists the way Nazis had capitalized on economic depression to capture power. Prosperity, then, would help curb Red-Fascist totalitarianism and its offspring, war.

Another component of the postwar outlook held that world prosperity, and hence peace, could be fostered only by healthy world trade. "A large volume of soundly based international trade is essential if we are to achieve prosperity in the United States, build a durable structure of world economy, and attain our goal of world peace and security," declared President Truman.[9] On another occasion he explained: "In fact the three— peace, freedom, and world trade—are inseparable."[10] Some commentators regretted that economic blocs, such as the British Sterling Bloc and the Soviet network in Eastern Europe, as well as bilateral treaties, impeded world trade. "Nations which act as enemies in the marketplace," Will Clayton remarked, "cannot long be friends at the council table."[11] Thus Studebaker's Paul Hoffman, later to head the Marshall Plan, applauded the $3.75 billion loan to Britain in 1946 and its provisions for relaxation of Sterling Bloc trade restrictions as "a first step toward international economic coordination. And international coordination is in turn a first step toward world peace."[12]

The United States in the postwar period attempted to combat trade restrictions through a multilateral, non-discriminatory, most-favored-nation trade policy. The concept of the "open door"—equal trade and investment opportunity—was considered by government spokespeople to be a traditional one from which the United States itself had strayed with a protectionist tariff policy and from which the United States still deviated in the postwar years through commodity agreements, domestic farm price supports, import quotas, subsidies to shipping interests and sugar producers, oil and other cartel agreements, and preferential trade relations with Cuba and the Philippines. Truman intended to effect the "open door" through multilateral trade agreements and the Reciprocal Trade Agreements program. Winthrop Aldrich, Chairman of the Chase National Bank and a frequent government adviser, summarized the American position: "Bilateralism is the road to war; multilateralism is the highway to peace."[13] In the terms for the General Agreement on Tariffs and Trade (GATT), the World Bank, and the International Trade Organization, American officials counter-attacked against restrictions on world trade.

Another element of the peace and prosperity concept was the interdependence of the world economy and thus the potential impact of an American depression on the world and a global depression on the United States. Henry Wallace noted that "full employment in the United States means prosperity all over the world."[14] Harriman pictured the United States as the "financial and economic pivot of the world," and agreed with the prevalent view "that economic stagnation in the United States would drag the rest of the world down with us."[15] American leaders were also alert to the potential effect of a world depression on the American economy. Truman found in 1947 that "our prosperity is endangered by hunger and cold in other lands," and his Secretary of the Interior, Julius Krug, remarked that "depressions are as catching as the common cold."[16]

A healthy and growing American trade was seen as the key to both world and American prosperity. Should that trade collapse, the spindly-legged world economy would tumble, leaving totalitarianism, war, and revolution behind. Postwar American foreign trade was a significant and perhaps critical element in the welfare of war-torn countries, because the United States was the largest supplier of goods to world markets. By 1947 American exports accounted for one-third of total world exports, one-fourth in 1948, and one-fifth in 1949, compared to one-seventh in 1936–38. And American direct investments abroad, amounting to $12 billion in 1950, provided precious capital and revenue-producing export goods for war-devastated European nations.

The United States' economic power was measured in other impressive figures. American interests owned or controlled 50 percent of total known world oil reserves in 1947. That same year American automobile production was eight times the output of France, England, and Germany combined, and four years later the United States turned out seven million cars, compared to Russia's 65,000. In the immediate postwar period, the United States was the largest producer and consumer of both coal and steel in the world. General Motors' Charles Wilson bragged in 1948 that, with only 6 percent of the world's area and 7 percent of its population, the United States had 46 percent of the world's electric power, 48 percent of its radios, 54 percent of its telephones, and 92 percent of its modern bathtubs. The economic power of American bathtubs may have been lost on foreigners, but not the strength of America's capital and durable goods manufacturing. Countries in dire need of economic rehabilitation could not ignore the conspicuous fact that in 1948 the United States produced over 40 percent of the world's goods and services, and accounted for almost one-half the world's industrial output. The President put it simply: "We are the giant of the economic world."[17]

The peace and prosperity concept held as well that foreign trade was vital to the economic power and prosperity of the United States. Willard L. Thorp, Assistant Secretary of State, declared that "any serious failure to maintain this flow [exports and imports] would put millions of American businessmen, farmers, and workers out of business."[18] Government and business executives amassed impressive statistics to demonstrate that foreign trade represented an important component of the economy. When the Secretary of the Agriculture presented his department's case for the Marshall Plan, he listed surpluses in citrus fruits, potatoes, eggs, dairy products, wheat, rice, cotton, and tobacco, all of which, he said, needed foreign markets. World food shortages and the American foreign aid and relief expenditures in the 1946–47 period insured large exports of American agricultural goods, although there was some friction between domestic demands and foreign needs. In 1947 almost half of the American production of wheat went abroad. Thirteen percent of total farm income was derived from exports. Although exports represented only about 7 percent of the gross national product and approximately 10 percent of the production of movable goods, certain key industries, such as automobiles and trucks, coal, machine tools, railroad locomotives, steel products, and farm machinery, relied heavily upon foreign trade for their well-being. In 1947, for example, exports of the Monsanto Chemical Company accounted for one-eighth of total sales, and General Motors exported about 10 percent of its sales in the postwar period.

Foreign trade was also important to employment. Both Henry Wallace and Eric Johnston, president of the United States Chamber of Commerce, were counting on foreign trade to contribute from four million to five million jobs at the close of the war. Actually, according to the Bureau of Labor Statistics, in the first half of 1947, 2,364,000 of 41,963,000 employees were directly or indirectly employed because of the export trade. These people accounted for 5.6 percent of all non-agricultural employees. In some industries, the percentage far exceeded this figure. In steel companies, for example, the figure was about 20 percent, in rubber 15 percent, in coal mining 18 percent, and in chemicals 11 percent.

Although these figures seemed small at first glance, Secretary of the Treasury John Snyder reminded the President that:

> The importance of U.S. exports to the American economy is evidenced by the fact that they exceed in volume such important single elements of the national product as expenditures on producers' durable equipment, consumers' expenditures on durable goods, the net change in business inventories, the total expenditures by State and local governments or even private construction.[19]

"Theoretically," wrote one observer in the *Harvard Business Review*, "the United States can achieve full employment without any exports whatsoever."[20] But such a condition would require grave changes in the American economy, as well as increased government controls. The Committee for Economic Development predicted a "great readjustment, much inefficient production, and a lower standard of living" for the United States were it to cut back its foreign trade.[21] Eric Johnston warned that a curtailment of American foreign trade "would mean vast population shifts, and . . . new ways of subsistence would have to be found for entire geographic regions."[22]

It was convincingly argued, too, that exports paid for the imports required by industry and the military. It became fashionable after the war to speak of the United States as a "have-not" nation because of the wartime drain on its resources. The director of the Bureau of Mines, for example, reported that the United States was self-sufficient in but a few of the most critical industrial minerals. Worried leaders itemized domestic deficiencies in zinc, tin, mercury, manganese, lead, cobalt, vanadium, tungsten, chromite, industrial diamonds, nickel, asbestos fiber, bauxite, petroleum, and copper, or a total of over fifty materials. President Truman was emphatic about the importance of raw materials to national security. "Without foreign trade . . . it would be difficult, if not impossible, for us to develop atomic energy."[23] Some forty materials from fifty-seven different countries were necessary for the production of steel. Various government programs, including the Marshall Plan and Point Four (see Chapter 8), attempted to purchase and stockpile vital raw materials.

Among other policymakers, Secretary Forrestal saw a direct link between foreign aid and the shortage in critical materials. For Forrestal, aid to Greece and Turkey under the Truman Doctrine was more than a simple effort to contain Communism; he labeled the doctrine "hard and selfish." What did he mean? Seventy-three percent of America's imports consisted of raw materials for the production of necessities for the United States, and 55 percent of these needed imports came from areas within the British Empire, mostly in Asia. Forrestal went on: "These raw materials have to come over the seas and a good many have to go through the Mediterranean. That is one reason why the Mediterranean must remain a free highway."[24] Again, Forrestal told one skeptical friend that the materials problem "is the only thing that makes any impression on me at all," and his listener observed that the "average fellow in this country" was not aware of the raw materials consideration.[25] Few Americans were aware either that the March 10 draft of the Truman Doctrine speech stated (later removed) that Greece and Turkey were areas of "great natural resources which must be accessible to all nations and must not be under the exclusive control or domination of any single nation. The weakening of Turkey, or the further weakening of Greece, would invite such control."[26] The materials question was a salient component of the peace and prosperity theme of an interdependent economic world and the need for foreign trade to protect American prosperity and security.

Both during and after World War II, many Americans and foreigners waited expectantly for the United States to suffer a postwar economic collapse. When the French Institute of Public Opinion asked in 1947 why the United States had proposed the Marshall Plan, over half of those who replied said that the program was designed to insure foreign markets for American goods, and thereby avert an American economic crisis. American leaders always protested this foreign assessment of their intentions and were quick to label it Communist propaganda. Indeed, the Soviet Union and the Communist parties throughout Europe did harp on this issue. Yet the idea that the United States gave foreign aid to head off a depression was not the creation of Russian propagandists, although they no doubt inflated it, for the belief was already widespread in the world and in the United States itself.

That prominent Americans feared economic decline—and the resultant dangers to American well-being, to survival of a world free from the entanglements of Communism, or, in brief, to a world at peace—was in large part a throwback to the depression of the 1930s. Would the United States, without wartime expenditures, resume the shaky economy it had suffered before the war intervened? Would the experience of boom-bust-quasi-boom in the 1919–21 period be repeated? Throughout 1945 and 1946 American observers differed on the business outlook, with those predicting a downturn having the edge. Amidst the pessimism, Secretary Snyder tried to lift spirits by declaring it "bad psychology to be continually talking about 'depression' and 'recession' when we have all the elements for a continued prosperous period."[27]

In early June 1947, the *Journal of Commerce* reported that "an international economic crisis is being freely predicted today. . . . In this country, an approaching collapse is ascribed to the exhaustion of purchasing power abroad."[28] By the middle of 1947 forecasts predicting recession were on the increase. Arousing fears were the economic distress in Europe and the administration's dire warnings at that time that continued European depression would have unfortunate domestic and international economic and political consequences unless programs such as the Truman Doctrine and the Marshall Plan were approved. There was mounting concern, too, over inflationary prices. Also playing a part was an attempt to use history. Some commentators compared the post-World War II period with the pattern of the early 1920s. Thus the middle of 1947 would coincide with the start of the recession in 1920, when a sharp decline in exports and a slowdown in construction had brought an economic slump.

Truman officials in mid-1947 did not argue, however, that European aid was essential to head off an imminent recession. Rather, they stressed the long-range economic and political importance of such assistance in the familiar peace and prosperity idiom. In November 1947 the Council of Economic Advisers predicted an $8 billion decline in exports unless a new foreign aid program was initiated. Although this rapid reduction would not "inflict serious short-run damage on our economy, substantial problems of readjustment would be generated. Moreover, the industrial paralysis which could be expected to result in some other countries would have repercussions of major proportion upon our own economy and upon world stability."[29] The Interim Aid program ($400 million to Austria, China, France, and Italy during the winter of 1947–48) and the Marshall Plan provided the dollars to help prevent such an occurrence, and continued a trend set in 1945–47, when foreign aid extended by the United States government financed one-third of American exports. In the second quarter of 1947, aid to Europe financed the United States' entire export surplus to that area.

The decision to expand the foreign aid program had a stimulating effect on Americans who had been pessimistic about their economic future. Machine-tool manufacturers welcomed the Marshall Plan as a much-needed vehicle for profit, and *Fortune* found business executives much more optimistic by November 1947. A White House memorandum of March 1948, looking back on the spring of 1947 when people expected an economic decline, considered the Marshall Plan "very important" in providing "a powerful psychological stimulus to a resumption of the upward spiral," because "people generally interpreted [it] as providing for a tremendous increase in the level of exports."[30] And wheat prices were kept up by relief shipments.

Yet the recession that came in the third quarter of 1948 and lasted until the second quarter of 1950 was not blocked by the foreign aid program. Exports continued to decline, and about one-fourth of the total drop in output during the recession was due to the slackening of the export trade. But foreign aid unquestionably blunted some of the sharper edges of the export decline and the recession by placing $6.45 billion in the hands of Europeans and other foreign people buying in American markets in 1948–49. It was mild, too, because, due to American aid, European business activity did not slacken at the same time.

Although many American leaders thought a decline in exports would deliver recession, this was not the only consideration in predicting an economic slump. Some people never mentioned the effect of a European crisis or reduced foreign trade. They often thought excessively high domestic prices—inflation—would produce the collapse. The president of Republic Steel, for example, blamed American generosity abroad for high prices at home. Other Americans feared that foreign aid plans would inflate the economy. Inflationary tendencies were at work in the economy, and Europe's demand for goods scarce in the United States did contribute to inflation. As the *Journal of Commerce* put it in the case of foreign aid, the country had been given "the choice . . . between higher prices or lower exports." Neither of the alternatives was welcome, but government officials and some labor and business leaders thought that at least prices could be controlled through anti-inflationary policies. When the President went before a special session of Congress in November 1947, he asked for two programs: economic controls to manage inflation and Interim Aid. Truman did not intend to make foreign aid contingent upon controls. Foreign aid and foreign trade were simply too important to both American and world prosperity, to the American effort to meet the Communist threat, and hence to world peace, to be permitted to fade. Secretary Marshall, in defending the program that bore his name against the charges that it would be inflationary, and in emphasizing the long-range concern, argued that inflation "would not create a situation for us that was serious compared to the troubles that we are trying to meet abroad."[31] Most American leaders agreed with the National Advisory Council on International Monetary and Financial Problems, which held that "any temporary sacrifice" caused in the American economy because of a foreign assistance program "will be small compared to the long-range advantages to the United States of a peaceful, active, and growing world economy."[32]

The words "long-range" are important to the peace and prosperity concept. The central question was not whether a decline in foreign trade would bring an immediate American recession or depression, but whether the interdependent world economy, linked by American exports and economic power, would be viable and prosperous enough to stave off unstable and radical politics—in short, Communism. This was the burden of the peace and prosperity theme expressed so forcefully in the Marshall Plan.

After European officials responded to the call in Marshall's address at Harvard University on June 5, 1947, for European initiative in a large-scale foreign aid plan, the Truman Administration began to cultivate public endorsement of a European Recovery Program (ERP). Standing with the almost solid front of business leadership in endorsing the Marshall Plan were the American Federation of Labor, the Congress of Industrial Organizations, Americans for Democratic Action, Veterans of Foreign Wars, and the American Farm Bureau Federation. Events in Europe spurred passage of the expensive Marshall Plan in early 1948. The dollar shortage there had become acute. The fear that the Communist Party would win the Italian elections in

April 1948 persuaded some to call for the use of foreign aid to influence the outcome. Soviet pressures on Finland and the Communist coup in Czechoslovakia in February also disturbed Congress. And by early 1948 it was clear and encouraging that coal-and-iron-rich West Germany would be included in the ERP and that the United States was trimming back its decartelization and industrial dismantling operations in the new economic partner. The Truman Administration exploited these events to create a crisis atmosphere. The President appeared before Congress in mid-March to plead alarmingly for temporary selective service, universal military training, and the Marshall Plan; and leaks from American military officials suggested administration fears of a serious and sudden conflict with the Soviet Union.

The administration's lobbying was helped by the Soviet Union's rejection of the Marshall Plan, thus assuring Americans that the plan would be an anti-Communist effort. In fact, the plan was presented in such a way that the Soviet Union and the Eastern European countries could not participate. Russian Foreign Minister V. M. Molotov had gone to the Paris meeting of European nations in June 1947 interested in Marshall's proposal, but the conferees sparred immediately. The Russians were suspicious of a private French-British meeting before the conference began, and the British were decidedly cool toward Molotov's delegation in Paris. Molotov asked the French Foreign Minister what Britain and France "had done behind his back."[33] Fearful that the United States would manipulate the Marshall Plan for diplomatic purposes, Molotov rejected the Anglo-French call for a Big Three steering committee to draw up country surveys. He responded that the sovereignty of each potential recipient had to be maintained, and that a Big Three committee, with American overlordship, might meddle in internal affairs. The Soviets balked at United States control over the recovery program, especially when the Marshall Plan followed so closely after the Truman Doctrine and anti-Soviet speeches by government officials like Dean Acheson, and after the United States had refused to grant a loan to the Russians in 1945–46 unless they accepted American trade principles and disengaged from Eastern Europe. Soviet power in Eastern Europe was neither dominant nor solidified in early 1947 (the Czech coup, for example, may have been a response to the Marshall Plan), and the massive influx of American-directed dollars certainly would have challenged Russian influence. The Soviets opposed as well the participation of its nemesis Germany in any recovery plan, because of their intense apprehension that a revived Germany would once again threaten Russia.

The Soviet Union was also wary of the French-British proposal, prepared before the Paris gathering, that the European economies be integrated. Russia interpreted this to mean that the *status quo antebellum* would be perpetuated; that is, that Western Europe would be the industrial center and Eastern Europe the supplier of raw materials, especially grains and coal. The Russians rejected the subordination of the agricultural East to the industrial West. The Eastern European countries had developed plans for industrialization. They were undeveloped countries, and their economic difficulties had been augmented by the war's destruction. The foreign affairs analyst Vera Micheles Dean wrote in 1948 that "experts on Eastern Europe believe that only through industrialization accompanied by modernization of agriculture can the countries of this region solve their rural overpopulation problems, and ultimately raise their standards of living."[34] The Marshall Plan would seemingly disrupt such reorientation.

Moscow, then, considered the Marshall Plan another American ploy to challenge and isolate Communist nations. The Soviet Union's response was to tighten its grip on Eastern Europe. The countries of the region—either because they did not wish to affront their Soviet neighbor by joining an American-directed plan or because the Soviets ordered them—did not join the ERP. In July 1947 Russia secured a number of trade agreements with its neighbors and eventually attempted to mold them into a "Molotov Plan." The International Communist Information Bureau (Cominform) was organized in September 1947 as the Russian propaganda agency whose main target was the ERP. In February 1948 Czechoslovakia was dragged into the Communist camp, after Jan Masaryk had tried desperately to steer a middle course between East and West and had failed to attract American foreign aid. The Marshall Plan, and the way it was presented, served to strengthen developing international blocs and to realize Secretary of State James Byrnes's simplistic division of the world in 1946 into "friends" and "enemies."[35]

Because the Marshall Plan was intended as a weapon of American foreign policy, the United States did not work through the United Nations or the Economic Commission for Europe (ECE). Perhaps the ECE was an alternative. That United Nations association was a working body which counted the Soviet Union as a member. Kennan had once suggested a program utilizing the ECE, and Walt W. Rostow, assistant to the ECE Executive Secretary, observed in 1949 that many world leaders saw the ECE as a means of reducing Soviet-American tensions. He pointed out that Russia did not disrupt the ECE, even after the Marshall speech. The Soviets cooperated in ECE functions, including its coal allocations. But the United States chose to form an organization it could control directly, and rejected the offer of United Nations Secretary General Trygve Lie for a program under ECE auspices.

Soviet participation in an American-directed Marshall Plan was highly unlikely, and American leaders did not expect Soviet acceptance. Moreover, it would have been the utmost of illogic and contradiction for Congress to approve funds for the Soviet Union so shortly after the passage of the anti-Soviet Truman Doctrine and the curtailment of United Nations Relief and Rehabilitation Administration supplies to Russia, and at a time when complaints were mounting against trade with the Soviets. Kennan advised Marshall to "play it straight," but Clayton emphasized that, although the Soviets would be invited, the United States "must run this show."[36] Truman officials were not gambling much in offering the Soviet Union a place in the Marshall Plan, for Moscow could not have accepted membership in an organization so conspicuously dominated by Americans and driven in large part by the objective of undermining Soviet influence. The American overture constituted a gesture and diplomatic finesse, placing the burden of rejection on Stalin.

The Marshall Plan proved a mixed success. By the end of 1952 the ERP nations had received $13 billion, and an additional $5 billion entered the treasuries of the European members of NATO. American exports to Western Europe were maintained, and vital raw materials were added to American stockpiles. In Europe, the ERP caused inflation, failed to solve a balance of payments problem, and took only tentative steps toward economic integration. But it energized Western European industrial production and investment and started the region toward self-sustaining economic growth. With the help of an extensive propaganda campaign, governments friendly to the United States were elected. American leaders credited the Marshall Plan with thwarting Communist electoral victories in Italy and France and Communist subversion elsewhere. Successful as it was, the ERP joined Soviet policies in dividing Europe. Stimulating as it was to allies' economies, it cannot be claimed that without the Marshall Plan the area would have succumbed to Communism. Traditional erratic politics and a slower economic pace, but not necessarily Communism, might have plagued Western Europe.

As the Marshall Plan showed, American foreign policy after World War II was more than simply "stop Communism" or "avert depression" or "expand markets and raw materials sources." The three objectives were intimately linked in the peace and prosperity concept, which explained the containment policy well and was its rationale. To the Truman Doctrine and the Marshall Plan was added NATO. Although Kennan considered NATO a perversion of his view of containment, other government officials argued that military bases and alliances were needed to achieve peace and prosperity. And, in symbiotic relationship, stable, prosperous economic conditions would serve American security by guaranteeing the supply of strategic raw materials, improving technology, strengthening allies, and providing the strong economic base needed for armed forces in the field. By 1952, American foreign aid had taken on a hefty military character: Eighty percent of the assistance to Western Europe consisted of military items.

Coupled with the shift to a military emphasis were certain weaknesses of the peace and prosperity notion itself. At a time when the United States was an economic behemoth, wealthy and expanding compared to the rest of the world, many foreigners interpreted the ardent American desire to extend United States foreign trade and investment as an attempt to dominate foreign markets and politics. Thus Henry Wallace's challenge to Stalin to enter into a production race with the United States must have seemed ridiculous to the Russian dictator, who was quite aware of the economic superiority of his wartime ally and the comparative weakness of his own war-wracked nation. In an October 10, 1946, speech, Molotov referred to the 1946 *World Almanac* to remind Americans how World War II had strengthened their economy. He asked what "equal opportunity" would really mean in Europe. How long, for example, he inquired, would Rumanian and Yugoslavian industry remain independent after the penetration of American-controlled capital? This was not simply Communist propaganda. Leopold S. Amery, British Conservative and former Secretary of State for India, observed: "American industry in all its phases is obsessed with the idea of conquering every market in the world. . . . They are out to capture aviation, sea transport and finance, the fields of invisible as well as visible exports."[37] And Republican Senator Robert Taft recognized that economic expansion might "build up hostility to us rather than any genuine friendship. It is easy to slip into an attitude of imperialism and to entertain the idea that we know what is good for other people better than they know themselves."[38] The president of the National Association of Wool Manufacturers did not share the foreign trade enthusiasm of much of the rest of the business community in 1947: "The emphasis which this country is currently putting on foreign trade is much more aptly characterized as 'economic aggression' than a means of promoting peace."[39] Few countries, in a time of distress, were eager to lay bare their economies to American goods with which they could not compete. Given international conditions and the way the United States went about trying to create prosperity, then, tension as much as "peace" flowed from the exertion of American economic power.

The peace and prosperity concept rested on the shaky and seldom demonstrated assumptions that economic improvement promoted peace, that foreign aid promoted stable politics or democracy, and that poverty produced Communism. Actually, the desperately poor, illiterate, and debilitated have seldom generated much Communist revolutionary activity. They are just as likely to produce non-Communist dictators—many of whom the United States has supported.

The peace and prosperity concept was not applied universally or consistently. In areas we now consider part of the Third World, marked by economic instability and a high degree of poverty, the United States tried to maintain an anti-Communist order through its military, political, and economic power— witness Latin America—or through the power of European colonial masters—witness Indochina and Africa. Often Washington seemed more committed to the creation of stable politics than to the eradication of those deplorable economic conditions that supposedly bred Communism. Peace and prosperity often meant political stability, not economic improvement, and support for colonialism to satisfy allies, not for national independence movements. Poverty may have bred Communism but so did colonialism. Moreover, dollars frequently shored up military or authoritarian regimes—witness Greece, Iran, Nicaragua, and the Philippines—rather than built up economies. The theme of peace and prosperity exaggerated the postwar Communist threat and induced an American expansion abroad that many foreign critics read as coercive or excessive. Americans preferred to call their anti-Communist mission containment. But, when it comes to expansion and containment, you can't have one without the other. President Truman came to know this well when he projected American power abroad.

# 3

# Harry S Truman, American Power, and the Soviet Threat

President Harry S Truman and his Secretary of State Dean Acheson, Henry A. Kissinger once remarked, "ushered in the most creative period in the history of American foreign policy."[1] Presidents from Eisenhower to Reagan have exalted Truman for his decisiveness and success in launching the Truman Doctrine, the Marshall Plan, and NATO, and for staring the Soviets down in Berlin during those hair-trigger days of the blockade and airlift. John F. Kennedy and Lyndon B. Johnson invoked memories of Truman and the containment doctrine again and again to explain American intervention in Vietnam. Jimmy Carter has written in his memoirs that Truman had served as his model— that he studied Truman's career more than that of any other president and came to admire greatly his courage, honesty, and willingness "to be unpopular if he believed his actions were the best for the country."[2] Some historians have gone so far as to claim that Truman saved humankind from World War III. On the other hand, he has drawn a diverse set of critics. The diplomat and analyst George F. Kennan, the journalist Walter Lippmann, the political scientist Hans Morgenthau, politicians of the left and right, like Henry A. Wallace and Robert A. Taft, and many historians have questioned Truman's penchant for his quick, simple answer, blunt, careless rhetoric, and facile analogies, his moralism that obscured the complexity of causation, his militarization of American foreign policy, his impatience with diplomacy itself, and his exaggeration of the Soviet threat.

Still, there is no denying the man and his contributions. He fashioned policies and doctrines that have guided leaders to this day. He helped initiate the nuclear age with his decisions to annihilate Hiroshima and Nagasaki with atomic bombs and to develop the hydrogen bomb. His reconstruction programs rehabilitated former enemies West Germany and Japan into thriving, industrial giants and close American allies. His administration's search for oil in Arab lands and endorsement of a new Jewish state in Palestine planted the United States in the Middle East as never before. Overall, Truman projected American power onto the world stage with unprecedented activity, expanding American interests worldwide, providing American solutions to problems afflicting countries far distant from the United States, establishing the United States as the pre-eminent nation in the postwar era.

Historians have given high marks to the President from Missouri with the memorable "give 'em hell Harry" style. In an elaborate polling of historians conducted in the early 1980s, Truman was judged "near great," just behind Andrew Jackson, Woodrow Wilson, and Theodore Roosevelt and just ahead of John Adams, Lyndon B. Johnson, and Dwight D. Eisenhower. He was also ranked seventh in a list of the most "controversial" Presidents—a list headed, of course, by Richard M. Nixon.[3] When historians distinguished Truman's attributes, they usually mentioned his activism as a "doer," foreign policy accomplishments, expansion of executive power, decisiveness, shaping and using of public opinion, and personal integrity.

On April 12, 1945, Vice President Truman was presiding over the United States Senate. He was bored, his thoughts wandering to a poker game scheduled that evening with friends at the Statler Hotel. Shortly after gaveling the Senate to adjournment that afternoon, Truman dropped into the private office of Speaker of the House Sam Rayburn to discuss some legislation and to strike a few liquid blows for liberty. Soon Truman learned that the White House had called: he should come over immediately and quietly. He put down his bourbon and water, apologized to Rayburn for the hurried departure, and hailed his chauffeur. Once in the White House Truman was escorted to the second floor study of Eleanor Roosevelt. There sad faces signaled Truman for the first time that something momentous was about to happen. Mrs. Roosevelt placed her hand on Truman's shoulder and announced that President Franklin D. Roosevelt had died. "Is there anything I can do for you?," asked a stunned Truman. Eleanor Roosevelt shook her head and replied: "Is there anything we can do for you? For you are the one in trouble now."[4]

Trouble indeed. As Truman confided to his diary that day, "the weight of the Government had fallen on my shoulders. . . . I knew the President had a great many meetings with Churchill and Stalin. I was not familiar with any of these things and it was really something to think about. . . ."[5] In fact, Truman as Vice President had never been included in high-level foreign policy discussions; between the inauguration and the President's death, Truman had met only three times with Roosevelt. And foreign affairs had never been a primary interest of Truman's; he had not sat, for example, on the Foreign Relations Committee during his ten years as a senator. Shortly after becoming President, Truman admitted to the Secretary of State that he "was very hazy about the Yalta [Conference] matters," especially about Poland.[6] Later he would lament that Roosevelt "never did talk to me confidentially about the war, or about foreign affairs or what he had in mind for the peace after the war."[7] The weight of foreign policy had fallen on him, and he knew so little. "I was plenty scared."[8] Apprehensive and insecure though he was, Truman was not content to sit in Roosevelt's shadow or brood about his inadequacies and new responsibilities. He would be "President in his own right," he told his first Cabinet meeting.[9] And through trial and error he became so.

About three months after assuming office, a more self-assertive but still self-doubting Truman boarded a ship for Europe, there to meet at Potsdam, near Berlin, with two of recent history's most imposing figures, Winston Churchill and Josef Stalin. "I sure dread this trip, worse than anything I've had to face," he wrote his beloved wife Bess.[10] On July 16, 1945,

Truman visited Berlin, where he witnessed the heavy destruction of the city, like much of Europe, now reduced to rubble. "I was thankful," he wrote later, "that the United States had been spared the most unbelievable devastation of this war."[11] At the Potsdam Conference Truman quickly took the measure of the eloquent Churchill and austere Stalin. "The boys say I gave them an earful," he boasted. He told his wife that "I reared up on my hind legs and told 'em where to get off and they got off."[12]

Truman's assertiveness at Potsdam on such issues as Poland and Germany stemmed not only from his forthright personality, but also from his learning that America's scientists had just successfully exploded an atomic bomb which could be used against Japan to end World War II. And more, it might serve as a diplomatic weapon to persuade others to behave according to American precepts. The news of the atomic test's success gave Truman "an entirely new feeling of confidence . . . ," Secretary of War Henry L. Stimson recorded in his diary. "Now I know what happened to Truman yesterday," commented Churchill. "When he got to the meeting after having read this report he was a changed man. He told the Russians just where they got off and generally bossed the whole meeting."[13]

Truman soon became known for what he himself called his "tough method." He crowed about giving Russia's Commissar for Foreign Affairs, V. M. Molotov, a "straight 'one-two to the jaw'" in their first meeting in the White House not long after Roosevelt's death.[14] Yet Secretary Stimson worried about the negative effects of Truman's "brutal frankness," and Ambassador Harriman was skeptical that the President's slam-bang manner worked to America's advantage.[15] Truman's brash, salty style suited his bent for the verbal brawl, but it ill-fit a world of diplomacy demanding quiet deliberation, thoughtful weighing of alternatives, patience, flexibility, and searching analysis of the motives and capabilities of others. If Truman "took 'em for a ride," as he bragged after Potsdam, the dangerous road upon which he raced led to the Cold War.[16] "It isn't any use kicking a tough hound [like the Russians] around because a tough hound will kick back," retired Secretary Cordell

Hull remarked after witnessing deteriorating Soviet-American relations.[17]

The United States entered the postwar period, then, with a new, inexperienced, yet bold President who was aware of America's enviable power in a world hobbled by war-wrought devastation and who shared the popular notion of "Red Fascism." To study this man and the power at his command, the state of the world in which he acted, his reading of the Soviet threat, and his declaration of the containment doctrine to meet the perceived threat further helps us to understand the origins of the Cold War. Truman's lasting legacy is his tremendous activism in extending American influence on a global scale—his building of an American "empire" or "hegemony." We can diagree over whether this postwar empire was created reluctantly, defensively, by invitation, or deliberately, by self-interested design. But few will deny that the drive to contain Communism fostered an exceptional, worldwide American expansion that produced empire and ultimately, and ironically, insecurity, for the more the United States expanded and drove in foreign stakes, the more vulnerable it seemed to become—the more exposed it became to a host of challenges from Communists and non-Communists alike.

In the aftermath of a war that bequeathed staggering human tragedy, rubble, and social and political chaos, "something new had to be created," recalled Dean Acheson. America's task "was one of fashioning, trying to help fashion what would come after the destruction of the old world."[18] World order had to be reconstructed; societies, political systems, and economies had to be rebuilt. Europe lost more than 30 million dead in the Second World War. Of this total, the Russians suffered between 15 and 20 million dead, the Poles 5.8 million dead, and the Germans 4.5 million dead. Asian casualties were also staggering: Japan lost 2 million, and millions of Chinese also died. Everywhere, armies had trampled farms and bombs had crumbled cities. Everywhere, transportation and communications sytems lay in ruins. Everywhere, water sources were contaminated and food supplies depleted. Everywhere, factories were gutted and lacked raw materials and labor. Everywhere, displaced persons searched for families and homes. One American journalist visited Warsaw, Poland, in 1945 and saw nothing but "rows of roofless, doorless, windowless walls. . . ."[19] An American general described Berlin as a "city of the dead."[20] In Greece one million people were homeless, agricultural production was down 50 percent, and 80 percent of railway rolling stock was inoperable. The bridges over the Danube River were demolished, and debris and bodies clogged the Rhine, Oder, and Elbe waterways. The Ukraine, once a center for coal, iron, steel, and farm goods, had been ravaged by the German scorched-earth policy. Much of the Soviet Union's national wealth had been destroyed. When the Secretary General of the United Nations visited Russia, he found "charred and twisted villages and cities . . . the most complete exhibit of destruction I have ever witnessed."[21]

To recount this grisly story of disaster is to emphasize that economic, social, and hence political "disintegration" characterized the postwar international system.[22] The question of how this disintegration could be reversed preoccupied Truman officials. Thinking in the peace and prosperity idiom, they believed that a failure to act would jeopardize American interests, drag the United States into depression and war, spawn totalitarianism and aggression, and permit the rise of Communists and other leftists who were eager to exploit the disorder. The prospects were grim, the precedents for action few, the necessities certain, the consequences of inaction grave. This formidable task of reconstruction drew the United States and the Soviet Union into conflict, for each had its own model for rebuilding states and each sought to align nations with its foreign policy.

Political turmoil within nations also drew America and Russia into conflict, for each saw gains to be made and losses to be suffered in the outcome of the political battles. Old regime leaders vied with leftists and other dissidents in state after state. In Poland, the Communist Lublin Poles dueled the conservative London Poles; in Greece the National Liberation Front contested the authority of the British-backed conservative Athens government; in China Mao Zedong's forces continued the civil war against Jiang Jieshi's (Chiang Kai-shek's) Nationalist regime. In occupied Germany, Austria, and Korea, the victors created competitive zones and backed different political groups. Much seemed at stake: economic ties, strategic bases, military allies, intelligence posts, and votes in international organizations. When the United States and the Soviet Union meddled in these politically unstable settings in their quest for influence, they collided—often fiercely.

The collapse of old empires also wrenched world affairs and invited confrontation between America and Russia. Weakened by the war and unable to sustain colonial armies in the field, the imperialists were forced to give way to nationalists who had long worked for independence. The British withdrew from India (and Pakistan) in 1947, from Ceylon and Burma the next year. The Dutch left Indonesia in 1949. The French clung to Indochina, engaged in bloody war, but departed in 1954. The European imperialists also pulled out of parts of Africa and the Middle East. The United States itself in 1946 granted independence to the Philippines. Decolonization produced a shifting of power within the international system and the emergence of new states whose allegiances both the Americans and Russians avidly sought.

With postwar economies, societies, politics, and empires shattered, President Truman confronted an awesome set of problems that would have bedeviled any leader. He also had impressive responsibilities and opportunities, because the United States had escaped from World War II not only intact but richer and stronger. America's abundant farmlands were spared from the tracks of marching armies, its cities were never leveled by bombs, and its factories remained in place. During the war, America's gross national product skyrocketed and every economic indicator, such as steel production, recorded significant growth (see Chapter 2). In the postwar years, Americans possessed the power, said Truman, "either to make the world economy work or, simply by failing to take the proper action, to allow it to collapse."[23] To create the American-oriented world the Truman Administration desired, and to isolate adversaries, the United States issued or withheld loans (giving one to Britain but not to Russia), launched major reconstruction programs like the Marshall Plan (see Chapter 2), and offered technical assistance through the Point Four Program (see Chapter 8). American dollars and votes also dominated the World Bank and International Monetary Fund, transforming them into instruments of American diplomacy.

The United States not only possessed the resources for reconstruction, but also the implements of destruction. The United States had the world's largest Navy, floating in two oceans, the most powerful Air Force, a shrinking yet still formidable Army, and a monopoly of the most frightening weapon of all, the atomic bomb. Not until after the Korean War did the United States stockpile many atomic bombs, but Secretary of State James F. Byrnes, like other American leaders, was known to say that he liked to use the atomic bomb for diplomatic leverage at conferences. Once, during a social occasion at the London Conference in the fall of 1945, Byrnes and Molotov bantered. The Soviet Commissar asked Byrnes if he had an atomic bomb in his "side pocket." Byrnes shot back that the weapon was actually in his "hip pocket." And, "if you don't cut out all this stalling and let us get down to work, I am going to pull an atomic bomb out of my hip pocket and let you have it."[24] Although Molotov apparently laughed, he could not have been amused, for he suspected that the Americans counted on the bomb as an implied threat to gain Soviet diplomatic concessions—and, as the supreme weapon, to blast the Soviet Union into smithereens in a war. Henry L. Stimson, for one, disapproved of "atomic diplomacy," because, he explained to the President, if Americans continued to have "this weapon rather ostentatiously on our hip, their [the Russians'] suspicions and their distrust of our purposes and motives will increase."[25]

Because of America's unusual postwar power, the Truman Administration could expand the United States sphere of influence beyond the Western Hemisphere and also intervene to protect American interests. But this begs a key question: Why did President Truman think it necessary to project American power abroad, to pursue an activist, global foreign policy unprecedented in United States history? The answer has several parts. First, Americans drew lessons from their experience in the 1930s. While indulging in their so-called "isolationism," they had watched economic depression spawn political extremism, which in turn, produced aggression and war. Never again, they vowed. No more appeasement with totalitarians, no more Munichs. "Red Fascism" became a popular phrase to express this American idea (see Chapter 1). The message seemed evident: To prevent a reincarnation of the 1930s, the United States would have to use its vast power to fight economic instability abroad. Americans felt compelled to project their power, second, because they feared, in the peace-and-prosperity thinking of the time, economic doom stemming from an economic sickness abroad that might spread to the United States, and from American dependency on overseas supplies of raw materials. To aid Europeans and other peoples would not only help them, but also sustain a high American standard of living and gain political friends, as in the case of Italy, where American foreign aid and advice influenced national elections and brought defeat to the left. The American fear of postwar shortages of petroleum also encouraged the Truman Administration to penetrate Middle Eastern oil in a major way. In Saudi Arabia, for example, Americans built and operated the strategically important Dhahran Airport and dominated that nation's oil resources.

Another reason why Truman projected American power so boldly derived from new strategic thinking. Because of the advent of the air age, travel across the world was shortened in time. Strategists spoke of the shrinkage of the globe. Places once deemed beyond American curiosity or interest now loomed important. Airplanes could travel great distances to deliver bombs. Powerful as it was, then, the United States also appeared vulnerable, especially to air attack. As General Carl A. Spaatz emphasized: "As top dog, America becomes target No. 1." He went on to argue that fast aircraft left no warning time for the United States. "The Pearl Harbor of a future war might well be Chicago, or Detroit, or even Washington."[26] To prevent such an occurrence, American leaders worked to acquire overseas bases in both the Pacific and Atlantic, thereby denying a potential enemy an attack route to the Western Hemisphere. Forward bases would also permit the United States to conduct offensive operations more effectively. The American strategic frontier had to be pushed outward. Thus the United States took the former Japanese-controlled Pacific islands of the Carolines,

Marshalls, and Marianas, maintained garrisons in Germany and Japan, and sent military missions to Iran, Turkey, Greece, Saudi Arabia, China, and to fourteen Latin American states. The Joint Chiefs of Staff and Department of State lists of desired foreign bases, and of sites where air transit rights were sought, included such far-flung spots as Algeria, India, French Indochina, New Zealand, Iceland, and the Azores. When asked where the American Navy would float, Navy Secretary James Forrestal replied: "Wherever there is a sea."[27] Today we may take the presumption of a global American presence for granted, but in Truman's day it was new, even radical thinking, especially after the "isolationist" 1930s.

These several explanations for American globalism suggest that the United States would have been an expansionist power whether or not the obstructionist Soviets were lurking about. That is, America's own needs—ideological, political, economic, strategic—encouraged such a projection of power. As the influential National Security Council Paper No. 68 (NSC-68) noted in April 1950, the "overall policy" of the United States was "designed to foster a world environment in which the American system can survive and flourish." This policy "we would probably pursue even if there were no Soviet threat."[28]

Americans, of course, did perceive a Soviet threat. Thus we turn to yet another explanation for the United States' dramatic extension of power early in the Cold War: to contain the Soviets. The Soviets unsettled Americans in so many ways. Their harsh Communist dogma and propagandistic slogans were not only monotonous; they also seemed threatening because of their call for world revolution and for the demise of capitalism. In the United Nations the Soviets cast vetoes and even on occasion walked out of the organization. At international conferences their *"nyets"* stung American ears. When they negotiated, the Soviets annoyed their interlocuters by repeating the same point over and over again, delaying meetings, or abruptly shifting positions. Truman labeled them "pigheaded," and Dean Acheson thought them so coarse and insulting that he once allowed that they were not "housebroken."[29]

The Soviet Union, moreover, had territorial ambitions, grabbing parts of Poland, Rumania, and Finland, and demanding parts of Turkey. In Eastern Europe, with their Red Army positioned to intimidate, the Soviets quickly manhandled the Poles and Rumanians. Communists in 1947 and 1948 seized power in Hungary and Czechoslovakia. Some Americans predicted that the Soviet military would roll across Western Europe. In general, Truman officials pictured the Soviet Union as an implaccable foe to an open world, an opportunistic nation that would probe for weak spots, exploit economic misery, snuff out individual freedom, and thwart self-determination. Americans thought the worst, some claiming that a Soviet-inspired international conspiracy insured perennial hostility and a creeping aggression aimed at American interests. To Truman and his advisers, the Soviets stood as the world's bully, and the very existence of this menacing bear necessitated an activist American foreign policy and an exertion of American power as a "counterforce."[30]

But Truman officials exaggerated the Soviet threat, imagining an adversary that never measured up to the galloping monster so often depicted by alarmist Americans. Even if the Soviets intended to dominate the world, or just Western Europe, they lacked the capabilities to do so. The Soviets had no foreign aid to dispense; outside Russia Communist parties were minorities; the Soviet economy was seriously crippled by the war; and the Soviet military suffered significant weaknesses. The Soviets lacked a modern navy, a strategic air force, the atomic bomb, and air defenses. Their wrecked economy could not support or supply an army in the field for very long, and their technology was antiquated. Their ground forces lacked motorized transportation, adequate equipment, and troop morale. A Soviet *blitzkrieg* invasion of Western Europe had little chance of success and would have proven suicidal for the Soviets, for even if they managed to gain temporary control of Western Europe by a military thrust, they could not strike the United States. So they would have to assume defensive positions and await crushing American attacks, probably including atomic bombings of Soviet Russia itself—plans for which existed.

Other evidence also suggests that a Soviet military threat to Western Europe was more myth than reality. The Soviet Union demobilized its forces after the war, dropping to about 2.9

million personnel in 1948. Many of its 175 divisions were
under-strength, and large numbers of them were engaged in
occupation duties, resisting challenges to Soviet authority in
Eastern Europe. American intelligence sources reported as well
that the Soviets could not count on troops of the occupied
countries, which were quite unreliable, if not rebellious. At
most, the Soviets had 700,000 to 800,000 troops available for an
attack against the West. To resist such an attack, the West had
about 800,000 troops, or approximate parity. For these reasons,
top American leaders did not expect a Soviet onslaught against
Western Europe. They and their intelligence sources emphasized Soviet military and economic weaknesses, not strengths,
Soviet hesitancy, not boldness.

Why then did Americans so fear the Soviets? Why did the
Central Intelligence Agency, the Joint Chiefs of Staff, and the
President exaggerate the Soviet threat? The first explanation is
that their intelligence estimates were just that—estimates. The
American intelligence community was still in a state of infancy,
hardly the well-developed system it would become in the 1950s
and 1960s. So Americans lacked complete assurance that their
figures on Soviet force deployment or armaments were accurate
or close to the mark. When leaders do not know, they tend to
assume the worst of an adversary's intentions and capabilities,
or to think that the Soviets might miscalculate, sparking a war
they did not want. In a chaotic world, the conception of a single,
inexorably aggressive adversary also brought a comforting sense
of knowing and consistency.

Truman officials also exaggerated the Soviet threat in order
"to extricate the United States from commitments and restraints
that were no longer considered desirable."[31] For example, they
loudly chastised the Soviets for violating the Yalta agreements;
yet Truman and his advisers knew the Yalta provisions were at
best vague and open to differing interpretations. But, more,
they purposefully misrepresented the Yalta agreement on the
vital question of the composition of the Polish government. In
so doing, they hoped to decrease the high degree of Communist
participation that the Yalta conferees had insured when they
stated that the new Polish regime would be formed by reor-
ganizing the provisional Lublin (Communist) government.

Through charges of Soviet malfeasance Washington sought to justify its own retreat from Yalta, such as its abandonment of the $20 billion reparations figure for Germany (half of which was supposed to go to the Soviet Union).

Another reason for the exaggeration: Truman liked things in black and white, as his aide Clark Clifford noted. Nuances, ambiguities, and counterevidence were often discounted to satisfy the President's preference for the simpler answer or his pre-conceived notions of Soviet aggressiveness. In mid-1946, for example, the Joint Chiefs of Staff deleted from a report to Truman a section that stressed Soviet weaknesses. American leaders also exaggerated the Soviet threat because it was useful in galvanizing and unifying American public opinion for an abandonment of recent and still lingering "isolationism" and support for an expansive foreign policy. Kennan quoted a colleague as saying that "if it [Soviet threat] had never existed, we would have had to invent it, to create a sense of urgency we need to bring us to the point of decisive action."[32] The military particularly overplayed the Soviet threat in order to persuade Congress to endorse larger defense budgets. This happened in 1948–49 with the creation of the North Atlantic Treaty Organization. NATO was established not to halt a Soviet military attack, because none was anticipated, but to give Europeans a psychological boost—a "will to resist."[33] American officials believed that the European Recovery Program would falter unless there was a "sense of security" to buttress it.[34] They nurtured apprehension, too, that some European nations might lean toward neutralism unless they were brought together under a security umbrella. NATO also seemed essential to help members resist internal subversion. The exaggerated, popular view that NATO was formed to deter a Soviet invasion of Western Europe by conventional forces stems, in part, from Truman's faulty recollection in his published memoirs.

Still another explanation for why Americans exaggerated the Soviet threat is found in their attention since the Bolshevik Revolution of 1917 to the utopian Communist goal of world revolution, confusing goals with actual behavior. Thus Americans believed that the sinister Soviets and their Communist allies would exploit postwar economic, social, and political disorder, not through a direct military thrust, but rather through covert subversion. The recovery of Germany and Japan became necessary, then, to deny the Communists political opportunities to thwart American plans for the integration of these former enemies into an American system of trade and defense. And because economic instability troubled so much of Eurasia, Communist gains through subversion might deny the United States strategic raw materials.

Why dwell on this question of the American exaggeration of the Soviet threat? Because it over-simplified international realities by under-estimating local conditions that might thwart Soviet/Communist successes and by over-estimating the Soviet ability to act. Because it encouraged the Soviets to fear encirclement and to enlarge their military establishment, thereby contributing to a dangerous weapons race. Because it led to indiscriminate globalism. Because it put a damper on diplomacy; American officials were hesitant to negotiate with an opponent variously described as malevolent, deceitful, and inhuman. They especially did not warm to negotiations when some critics were ready to cry that diplomacy, which could produce compromises, was evidence in itself of softness toward Communism.

Exaggeration of the threat also led Americans to misinterpret events and in so doing to prompt the Soviets to make decisions contrary to American wishes. For example, the Soviet presence in Eastern Europe, once considered a simple question of the Soviets' building an iron curtain or bloc after the war, is now seen by historians in more complex terms. The Soviets did not seem to have a master plan for the region and followed different policies in different countries. Poland and Rumania were subjugated right away; Yugoslavia, on the other hand, was an independent Communist state led by Josip Tito, who broke dramatically with Stalin in 1948; Hungary conducted elections in the fall of 1945 (the Communists got only 17 percent of the vote) and did not suffer a Communist coup until 1947; in Czechoslovakia, free elections in May 1946 produced a non-Communist government that functioned until 1948; Finland, although under Soviet scrutiny, affirmed its independence. The Soviets did not have a firm grip on Eastern Europe before 1948—a prime reason why many American leaders believed the Soviets harbored weaknesses.

American policies were designed to roll the Soviets back. The United States reconstruction loan policy, encouragement of dissident groups, and appeal for free elections alarmed Moscow, contributing to a Soviet push to secure the area. The issue of free elections illustrates the point. Such a call was consistent with cherished American principle. But in the context of Eastern Europe and the Cold War, problems arose. First, Americans conspicuously followed a double standard which foreigners noted time and again; that is, if the principle of free elections really mattered, why not hold such elections in the United States' sphere of influence in Latin America, where an unsavory lot of dictators ruled? Second, free elections would have produced victories for anti-Soviet groups. Such results could only unsettle the Soviets and invite them to intervene to protect their interests in neighboring states—just as the United States had intervened in Cuba and Mexico in the twentieth century when hostile groups assumed power. In Hungary, for example, it was the non-Communist leader Ferenc Nagy who delayed elections in late 1946 because he knew the Communist Party would lose badly, thereby possibly triggering a repressive Soviet response. And, third, the United States had so little influence in Eastern Europe that it had no way of insuring free elections—no way of backing up its demands with power.

Walter Lippmann, among others, thought that the United States should tame its meddling in the region and make the best out of a bad arrangement of power. "I do believe," he said in 1947, "we shall have to recognize the principle of boundaries of spheres of influence which either side will not cross and have to proceed on the old principle that a good fence makes good neighbors."[35] Kennan shared this view, as did one State Department official who argued that the United States was incapable of becoming a successful watchdog in Eastern Europe. American "barkings, growlings, snappings, and occasional bitings," Cloyce K. Huston prophesized, would only irritate the Soviets without reducing their power.[36] Better still, argued some analysts, if the United States tempered its ventures into European affairs, then the Soviets, surely less alarmed, might tolerate more openness. But the United States did not stay out. Americans tried to project their power into a region where they had little chance of succeeding, but had substantial opportunity to irritate and alarm the always suspicious Soviets. In this way, it has been suggested, the United States itself helped pull down the iron curtain.

Another example of the exaggeration of the Soviet threat at work is found in the Truman Doctrine of 1947. Greece was beset by civil war, and the British could no longer fund a war against Communist-led insurgents who had a considerable non-Communist following. On March 12, Truman enunciated a universal doctrine: It "must be the policy of the United States to support free peoples who are resisting attempted subjugation by armed minorities or by outside pressures."[37] Although he never mentioned the Soviet Union by name, his juxtaposition of words like "democratic" and "totalitarian" and his references to Eastern Europe made the menace to Greece appear to be the Soviets. But there was and is no evidence of Soviet involvement in the Greek civil war. In fact, the Soviets had urged both the Greek Communists and their allies the Yugoslavs to stop the fighting for fear that the conflict would draw the United States into the Mediterranean. And the Greek Communists were strong nationalists. The United States nonetheless intervened in a major way in Greek affairs, becoming responsible for right-wing repression and a military establishment that plagued Greek politics through much of its postwar history. As for Turkey, official Washington did not expect the Soviet Union to strike militarily against that bordering nation. The Soviets were too weak in 1947 to undertake such a major operation, and they were asking for joint control of the Dardanelles largely for defense, for security. Then why did the President, in the Truman Doctrine speech, suggest that Turkey was imminently threatened? American strategists worried that Russia's long-term objective was the subjugation of its neighbor. But they also wished to exploit an opportunity to enhance the American military position in the Mediterranean region and in a state bordering the Soviet Union. The Greek crisis and the Truman Doctrine speech provided an appropriate environment to build up an American military presence in the Eastern Mediterranean for use against the Soviets should the unwanted war ever come. Truman's alarmist language further fixed the mistaken idea in the American mind that the Soviets were unrelenting aggressors intent upon undermining peace, and that the United States, almost alone, had to meet them everywhere. Truman's exaggerations and his commitment to the containment doctrine did not go unchallenged. Secretary Marshall himself was startled by the President's muscular anti-Communist rhetoric, and he questioned the wisdom of overstating the case. The Soviet specialist Llewellyn Thompson urged "caution" in swinging too far toward "outright opposition to Russia. . . ."[38] Walter Lippmann, in reacting to both Truman's speech and George F. Kennan's now famous "Mr. 'X'" article in the July 1947 issue of the journal *Foreign Affairs*, labeled containment a "strategic monstrosity," because it made no distinctions between important or vital and not-so-important or peripheral areas. Because American power was not omnipresent, Lippmann further argued, the "policy can be implemented only by recruiting, subsidizing and supporting a heterogeneous array of satellites, clients, dependents and puppets." He also criticized the containment doctrine for placing more emphasis on confrontation than on diplomacy.[39]

Truman himself came to see that there were dangers in stating imprecise, universal doctrines. He became boxed by his own rhetoric. When Mao Zedong's forces claimed victory in 1949 over Jiang's regime, conservative Republicans, angry Democrats, and various McCarthyites pilloried the President for letting China "fall" (see Chapter 4). China lost itself, he retorted. But his critics pressed the point: if containment was to be applied everywhere, as the President had said in the Truman Doctrine, why not China? Truman appeared inconsistent, when, in fact, in the case of China, he was ultimately prudent in cutting American losses where the United States proved incapable of reaching its goals. Unable to disarm his detractors on this issue, Truman stood vulnerable in the early 1950s to political demagogues who fueled McCarthyism. The long-term consequences in this example have been grave. Democrats believed they could never lose "another China"—never permit Communists or Marxists, whether or not linked to Moscow, to assume power abroad. President John F. Kennedy later said, for example, that he could not withdraw from Vietnam because that might be perceived as "another China" and spark charges that he was soft on Communism. America, in fact, could not bring itself to open diplomatic relations with the People's Republic of China until 1979.

Jiang's collapse joined the Soviet explosion of an atomic bomb, the formation of the German Democratic Republic (East Germany), and the Sino-Soviet Friendship Treaty to arouse American feeling in late 1949 and early 1950 that the Soviet threat had dramatically escalated. Although Kennan told his State Department colleagues that such feeling was "largely of our own making" rather than an accurate accounting of Soviet actions, the composers of NSC-68 preferred to dwell on a more dangerous Soviet menace in extreme rhetoric not usually found in a secret report.[40] But because the April 1950 document was aimed at President Truman, we can certainly understand why its language was hyperbolic. The fanatical and militant Soviets, concluded NSC-68, were seeking to impose "absolute authority over the rest of the world." America had to frustrate the global "design" of the "evil men" of the Kremlin, who were unrelentingly bent on "piecemeal aggression" against the "free world" through military force, infiltration, and intimidation.[41] The report called for a huge American and allied military build-up and nuclear arms development.

NSC-68, most scholars agree, was a flawed, even amateurish document. It assumed a Communist monolith that did not exist, drew alarmist conclusions based upon vague and inaccurate information about Soviet capabilities, made grand, unsubstantiated claims about Soviet intentions, glossed over the presence of many non-democratic countries in the "free world," and recommended against negotiations with Moscow at the very time the Soviets were advancing toward a policy of "peaceful co-existence." One State Department expert on the Soviet Union, Charles E. Bohlen, although generally happy with the report's conclusions, faulted NSC-68 for assuming a Soviet plot for world conquest—for "oversimplifying the problem." No, he advised, the Soviets sought foremostly to maintain their regime and to extend it abroad "to the degree that is possible without serious risk to the internal regime."[42] In short, there were limits to Soviet behavior. But few were listening to such cautionary voices. NSC-68 became American dogma, especially when the outbreak of the Korean War in June of 1950 sanctified it as a prophetic "we told you so."

The story of Truman's foreign policy is basically an accounting of how the United States, because of its own expansionism and exaggeration of the Soviet threat, became a global power. Truman projected American power after the Second World War to rehabilitate Western Europe, secure new allies, guarantee strategic and economic links, and block Communist or Soviet influence. He firmly implanted the image of the Soviets as relentless, worldwide transgressors with whom it is futile to negotiate. Through his exaggeration of the Soviet threat, Truman made it very likely that the United States would continue to practice global interventionism years after he left the White House.

# 4

## If Europe, Why Not China?
## Confronting Communism in Asia

"In Europe we were playing with good stuff," Dean Acheson remembered two years after leaving his secretaryship. "When you went ahead, you helped people who were willing, who had the will to be helped. The great trouble in China was that there wasn't that will." Like Harry S Truman, he guessed that one million American soldiers would have been needed to "save" China from the Communists. Once in, "how to let go of the thing?" For years, Washington would have to impose reforms on China's government and spend a colossal amount of money. "I don't know if the American people would have taken on a task like that." And, "would we have ended up being enemy foreigners? There would have been Communist propaganda all over the place. No one wants to be governed." Acheson concluded in defense of Truman Administration policy that "there are too many people in China and too little arable land, while the damned Americans ride around in Cadillacs—Oh, you would have one hell of a time."[1]

On March 18, 1947, when he was Acting Secretary of State, Acheson spoke less forthrightly. The initial question addressed to him in the State Department's first formal press conference after President Truman announced his doctrine was blunt: "What difference is there between the situation in China and

54

Greece which leads us to help one put down Communism and help the other to bring them into government?" Acheson shot back: "Is that a question which is asked to try to lure me into trouble, or are you really looking for information?" No doubt grumbles rippled through the assembled journalists; they had seldom enjoyed comfortable relations with the often stiff and overbearing Acheson. Although the correspondent assured Acheson that he was seeking information, not trouble, the diplomat dodged this query. He strangely said that he would "talk about that with you privately," then began to tell a second-hand story, which hardly clarified the issue. Once upon a time, Acheson remarked, his old friend and law associate Judge Covington was feasting at an oyster roast on the Eastern Shore. When the oysters were brought in, an elderly politician immediately grabbed a piping hot one, put it in his mouth, and quickly spat it out on the carpet, exclaiming: "A damn fool would have swallowed that."[2]

Acheson's reluctance to handle the delicate question was typical of the Truman Administration's hesitancy before 1949 to speak frankly when comparing American policies and aid programs for civil-war-wracked China on the one hand with those for an economically hobbled Europe on the other. In public, the administration did not adequately explain, after the announcement of the Truman Doctrine and the publication of George F. Kennan's "X" article in *Foreign Affairs*, both counseling containment on a global scale, why the United States did not commit itself to major activity in China to match that undertaken for Europe. Compared to the energetic American programs in Europe, which promised billions of dollars in aid to Greece and Turkey, post-UNRRA assistance, Interim Aid, the Marshall Plan, and NATO, and around which the United States built a protective economic and military shield, United States military and economic projects for China, while expensive, seemed hesitant and uncoordinated, given the perceived threat of a Communist victory. And until 1947, through mediation efforts like those of the Marshall Mission, the United States seemed to be inviting the Chinese Communists into a coalition government, whereas in Europe it was seeking to isolate them politically. The administration thus left itself vulnerable to the charge of inconsistency in implementing containment, the central principle of American Cold War diplomacy. "If Europe, why not China?" demanded the critics.

Criticism of America's China policy actually began before Truman enunciated his doctrine in March 1947. Ambassador to China Patrick Hurley resigned his post in November 1945 amid a storm of protest against "professional diplomats" who, he charged, "sided with the Chinese Communist armed party."[3] Early in 1946 the administration released the secret text of the Yalta agreements on the Far East. A flurry of accusations bombarded officials in Washington, with the "Manchurian Manifesto" of May 1946 signed by Congressman Walter Judd, Mrs. Clare Booth Luce, and Henry Luce, among others, complaining that the agreements reached at Yalta, granting territory in Asia to the Soviet Union, were "made behind China's back. . . . "[4] The notion gained vague currency that the United States had sold out Nationalist Jiang Jieshi (Chiang Kai-shek) to the Communists of Mao Zedung (Mao Tse-tung), and that the Truman Administration was simply not paying enough attention to China. Senator Arthur H. Vandenberg mused in January 1947, "we might just as well begin to face the Communist challenge on every front"—not just in Europe or Iran but also in China.[5] There were complaints, too, about the Marshall Mission of 1945–47. General George C. Marshall had attempted to create a unified government of Jiang Nationalists and Mao Communists, but he had failed. Critics were uneasy about Marshall's design to bring the Communists into a new Chinese government because, as Vandenberg asserted: "I never knew a Communist to enter a coalition government for any other purpose than to destroy it."[6]

The unveiling of the Truman Doctrine helped focus American concern about Communism in China directly on methods the United States could use to contain the danger. At a White House meeting two days before his speech, the President briefed leading congressmen on his forthcoming request for aid to Greece and Turkey. Vandenberg told him then that China was as important as Greece. Three months later the Michigan senator informed Marshall, now Secretary of State, that Congress wanted to see the total balance sheet for the distribution of

American aid. Where, in short, would the containment line be drawn? In the fall 1947 hearings for interim aid to Europe, Vandenberg again reminded Marshall that an assistance package that excluded China would be considered by some congressmen as "just a one-legged program."[7]

A stalwart of the so-called China Lobby also hammered on the issue of implementing containment in both Europe and China. Walter Judd, Republican congressman from Minnesota and self-professed expert on China affairs, wanted to make sure that "the Russians do not eliminate one front and become able to concentrate all their attention on the other—Europe."[8] He complained that the United States was spending billions on the European flank, but little or nothing on the other flank, Asia. Judd liked to use a familiar analogy: Communism had to be stopped at first base, China, or it would get to second base, Asia (including Japan), then to third base, Africa, and finally across home plate in Western Europe and the United States itself. Governor Thomas E. Dewey of New York, in a November 1947 speech prepared in part by Republican congressmen, sketched another picture: The world was a patient with gangrene in both legs, Europe and Asia. The patient could not survive, the soon-to-be Republican candidate for President warned, by saving only one leg. General Douglas MacArthur took time from administrative duties in Japan to inform the House Foreign Affairs Committee that the "Chinese problem is part of a global situation. . . . Fragmentary decisions in disconnected sectors of the world will not bring an integrated solution."[9] Republican Senator William Knowland added that "it did not make sense to try to keep 240,000,000 Europeans from being taken behind the iron curtain while we are complacent and unconcerned about 450,000,000 Chinese going the same way. . . ."[10]

Did the United States, as the critics charged, neglect the Chinese "front," "leg," or "flank" while concentrating on Europe? The criticism voiced by the Judds and Knowlands did not accord with the realities of American Cold War diplomacy. In postwar Asia the United States assumed new obligations, made influential decisions, and operated programs which belied the critics' charge that the Truman Administration slighted that region. Unabashedly the United States took control of former

Japanese islands in the Pacific to use them as military bases, and American officials administered the Ryukyu Islands. American troops were stationed in South Korea until 1949. The Philippines gained independence in 1946 but remained an American bastion and economic satellite with the help of the Philippines Rehabilitation Act of 1946, which provided for $520 million in aid and the transfer of $100 million worth of American surplus property to the new government. In Southeast Asia, American acquiescence in the re-establishment of French colonial rule in Indochina helped determine the history of that area. In the Dutch East Indies, American pressure and advice were pivotal in mediating between the Dutch and Indonesians, leading to independence for the Republic of Indonesia in 1949 and creation of a non-Communist nationalist government. And under the tutelage of General MacArthur, Japan was rebuilt as an ally, especially after 1948 when it became evident that China was going Communist.

As for China itself, American assistance in the period 1945 to 1949 was substantial—over three billion dollars, continuing the aid programs that had begun in World War II. Indeed, the United States became involved in the Chinese civil war overwhelmingly on the side of Jiang's Nationalist regime. President Truman was exuberant, if premature, when he remarked at a cabinet meeting in August 1946: "For the first time we now have a voice in China and for the first time we will be in a position to carry out the policy of 1898."[11] Once begun, American assistance to Jiang built up a momentum that was difficult to reverse. Marshall told a State-War-Navy Coordinating Committee meeting in mid-1947, for example, during a discussion about supplying the Nationalist army with more ammunition, that "it appears that we have a moral obligation to provide it inasmuch as we aided in equipping it with American arms."[12]

During the war, China received $846 million in Lend-Lease supplies; in 1942 Jiang obtained a $500 million stabilization credit; from 1944 to 1947 the United Nations Relief and Rehabilitation Administration spent $670 million in China, about three-quarters of the amount being American in origin. At the end of the war, Americans equipped and helped train thirty-nine Chinese divisions and United States planes transported one-half million Nationalist troops to North China and Manchuria. American Marines occupied North China until mid-1947 (peak strength 55,000), and the U.S. Army Advisory Group and Naval Advisory Group, numbering over fourteen hundred in mid-1948, were assigned to aid Jiang's forces. In the postwar period, China received a total of $769 million in Lend-Lease aid; in June 1946 Washington granted China a $50 million loan for "pipeline" goods. In 1946 and 1947, some 131 American naval vessels valued at $140 million were transferred to China; excess Army stocks in West China went to Jiang's regime in 1945 under an American loan of $20 million; in 1945 and 1946, United States Navy ordnance supplies worth $17.7 million also became Nationalist property. In the summer of 1947, when United States Marines retired from North China, 6,500 tons of American ammunition were left for Jiang's forces. In sum, American surplus property with a value of more than one billion dollars had been shifted to Jiang by early 1949. China also received approximately $44 million in post-United Nations Relief and Rehabilitation Administration assistance. In 1945 and 1946 the Export-Import Bank authorized six credits totaling another $69 million for exports to China of American cotton, ships, railway, power plant, and coal mining equipment. Early in 1946, the Bank also designated $500 million as additional credits for China, but the earmarked amount expired in mid-1947. In the China Aid Act of April 1948, Congress authorized $338 million in economic assistance ($275 million was appropriated) and another $125 million primarily for military purposes. And the Mutual Defense Assistance Act of 1949 provided $75 million for China.

Secretary Acheson claimed in early 1949 before a congressional committee that the United States had done more for China than for any other country in the postwar period. He insisted, drawing upon the reports of General David G. Barr, head of the military advisory group, that not one engagement in China had been lost through the lack of equipment or military supplies. Mao had angrily complained that the United States, active in the civil war, "supplies the money and guns and Chiang Kai-shek the men to fight for the United States and slaughter the Chinese people. . . . "[13] The Communist leader would have been more accurate had he noted, as did the Foreign Service Officer O. Edmund Clubb, that the "American arms supply had sufficed for *both* sides," because large caches of American military equipment fell into the hands of Communist forces.[14]

In the period 1945 to 1949, the United States became Jiang's major ally and chief quartermaster in the civil war. At the same time, the Truman Administration urged the Nationalist government to undertake political and economic reforms. From August 1946 to May 1947, Washington actually curtailed exports of munitions to China, although items already contracted were shipped. Washington seldom met Jiang's requests for aid in full, deciding instead on conditional support. American officials constantly pressed the Guomindong (Kuomintang) leader to form a more representative regime and to ferret out corruption from his government. "There are few harder stunts of statesmanship," Herbert Feis has written, "than at one and the same time to sustain a foreign government and to alter it against its fears and inclinations."[15] Failing in its reform thrusts, the United States still clung to the faltering Nationalist regime, feeding and fueling its armies and politicians. The perception of a Soviet-Communist threat against China and the postwar United States commitment to the containment of such a threat help explain why the United States, however reluctantly, backed and sustained Jiang—and ultimately went down to defeat with him.

After World War II American leaders perceived a Soviet-Communist danger in Asia, particularly in China. Although they recognized that the civil war in China had indigenous roots, they believed its consequences carried global implications. They came to interpret Chinese events through a Cold War lens; they foresaw an ultimate Soviet-American contest for influence in China. American policy was designed to deny the Soviets that influence and to preserve the traditional Open Door policy, which insured some American influence. Continued political and economic instability or Communist victory in the civil war might permit the Soviets into China. If this occurred, it was feared, Russia could use China as a springboard to Communize, through infiltration, the rest of Asia; threaten Ameri-

ca's ally, Japan; close trade routes and commercial opportunities; block Western access to such raw materials as tin and
rubber; and in the event of Soviet-American war, launch military operations from the vast Chinese mainland.

In the immediate postwar period, the Soviet threat against
China was often explained as indirect, quiescent, and long-
term, but ever present. In the American mind, the threat
became more ominous as the Chinese Communists unrelentingly threw defeat at Jiang. The United States had to make a
strong stand in China, declared Truman in 1945, or Russia
would take the place of Japan in the Far East. General Marshall
reported from China early in 1946 that he had pressed the
Nationalists to build a unified government "at the fastest
possible pace so as to eliminate her present vulnerability to
Soviet undercover attack, which exists so long as there remains
a separate Communist government and a separate Communist
army in China."[16] In his influential "long telegram" of February
22, 1946, George F. Kennan included the Chinese Communists
as a group willing to lend itself to "Soviet purposes."[17] Acheson
spoke of Russia's "predatory aims in Asia" and argued at an
August 1946 cabinet meeting that United States Marines should
remain in China, for "we will prevent by our very presence and
by the presence of our Marines some other country from
interfering in China to our regret."[18] The other country was, of
course, Russia.

Soviet behavior in Manchuria throughout 1945 and 1946
confirmed fears of a Soviet thrust. Despite American protests,
the Soviets thwarted Nationalist authority over the province,
seized war booty in the form of food and industrial equipment,
and, when they departed Manchuria in April, turned over large
stocks of weapons to Mao's Communist forces. The American
ambassador to China, John Leighton Stuart, expressed a popular assumption when he reported that Soviet favoritism toward
the Chinese Communists "fits customary Soviet predilection for
indirect activity wherever possible."[19] American officials reported after 1946 that they could not detect direct Soviet
involvement in the civil war. Still, they saw Russia hovering like
a menacing giant over the Chinese tumult, waiting to take
advantage of chaos and American weakness. An American

withdrawal from China, the Asian expert John Carter Vincent said, would open the door to Soviet influence. Thus there was no intention to abandon or "wash our hands" of the China problem, he assured the Chinese Ambassador to the United States in August 1946.[20]

The Joint Chiefs of Staff (JCS) were more emphatic than most officials in 1947 that the Soviets were plotting to master China. Taking the prevalent global perspective as expressed in the Truman Doctrine, they concluded that "in China, as in Europe and in the Middle East and Far East, it is clearly Soviet policy to expand control and influence wherever possible."[21] Vincent dissented from their call for increased military aid to Jiang and questioned not so much the Soviet threat, but its schedule: "A USSR-dominated China is not a danger of sufficient immediacy or probability" to justify large-scale intervention.[22] Still, Vincent seems to have been in the minority in minimizing the Soviet danger, and he was removed from the debate in July 1947 when he was sent off to Switzerland to be the American minister there. By November 1947 Secretary Marshall summarized American policy as preventing the "Soviet domination of China."[23]

American officials increasingly saw the Chinese Communists as tools, instruments, or vehicles for exercising Soviet influence. Throughout 1948 and 1949, as the Cold War intensified in Europe, American officials expanded upon their earlier feelings about the Soviet threat, with growing emphasis on the linkage between Mao and Moscow. Stuart beat the theme constantly in his cables to Washington. He bemoaned Mao's parroting of Soviet theories, Chinese Communist ideological subservience to Moscow, the Chinese Communist Party's aping of the Soviet model in dealing with foreign correspondents, and, in general, Mao's commitment to international Communism. Mao, concluded the Ambassador, was a "brilliant disciplined if somewhat junior ally in [the] 'world anti-imperialist front'. . . . "[24] In September of 1948 the Policy Planning Staff prepared a report, ultimately circulated as National Security Council Paper No. 34, which reflected the thinking of the Truman Administration. The report identified the USSR's covetous designs on Manchuria and North China and asserted: "It is the political situation in China which must arouse the aggressive interest of the

Kremlin."[25] The Policy Planning Staff warned that from China, Russia could mount a political offensive against the rest of East Asia. In short, Russia's objective was to control all of the territory comprising China.

Thus, by July 1949, when the Truman Administration published the justification for its post-1945 policies in a White Paper, it was not surprising to read that the Chinese "Communist leaders have foresworn their Chinese heritage and have publicly announced their subservience to a foreign power, Russia. . ." or that the Chinese Communist government served the interests of Soviet Russia.[26] By March of the following year Acheson was speaking about Soviet-Communist imperialism in Asia as part of the Kremlin's attempt to extend its absolute domination over the widest possible areas of the world.

Throughout these years, there was lingering speculation that the Chinese Communists would resist Soviet intrusion upon Chinese sovereignty. Americans hoped that Mao would become an Asian Tito, practicing a brand of independent Communism like that of Yugoslavia. Some American officials commented upon Titoist tendencies, but were uncertain whether the immediate examples of Sino-Soviet tension portended a long-term split. The Central Intelligence Agency suggested in November 1948 that "Chinese nationalism might well prove stronger than international communism" and that "their potential independence of the Kremlin is greater than that of Tito and, except for him, unique."[27] The Consul General in Shanghai, John M. Cabot, was more convinced than most Americans that China would go the way of Yugoslavia and that Russia would fail to control the independent-minded Chinese. He pointed out to the Secretary of State that the Chinese Communists were even less beholden than Tito to the Soviets for their assumption of power; that the Chinese were not committed to the Soviet economic bloc; that the Chinese had traditional links with the West; that Russia was historically an imperialist threat to China; and that Chinese Communism hardly resembled Soviet Communism.

In the White Paper Acheson noted the possibility that in the future Sino-Soviet relations might sour. After boldly declaring China subservient to Russia, he remarked that "ultimately the profound civilization and the democratic individualism of China will reassert themselves and she will throw off the foreign yoke."[28] National Security Council Paper No. 48, prepared in December 1949, after Jiang's fall, provided extensive discussion of this possibility. Like so many other American analyses, it identified Russia's aggressive intentions in Asia and its domination of the Chinese Communist Party. But it suggested that serious friction might develop if Russia tried to establish control over China similar to that of the subjugated Eastern European states, and concluded that American policy should exploit rifts between Communist China and Soviet Russia. Acheson told the Joint Chiefs of Staff, a few days after the paper circulated, that he perceived inevitable conflict between Beijing (Peking) and Moscow: "Mao is not a true satellite in that he came to power by his own efforts and was not installed in office by the Soviet Army."[29] American leaders thus seemed aware of a possible means of implementing containment—encourage a Sino-Soviet split. The United States goal remained what it had been since the end of the Second World War—to contain Soviet influence in China. Although remarking on a potential Stalin-Mao split, Truman Administration officials nonetheless took no steps to encourage a schism. Instead, they repeatedly pointed to the Soviet menace in the guise of Chinese Communists and assisted Jiang to the moment of his defeat. The President's comment to a January 1949 Cabinet meeting that "we can't be in a position of making any deal with a Communist regime" represented the core of American policy, even in the late fall 1949 through early spring 1950, after Jiang's collapse on the mainland, when American officials may have been warming to the idea of some degree of accommodation with Mao's government.[30]

China Lobbyists like Congressman Judd and Senator Styles Bridges asked, if the Truman Administration admitted that Communism directed by Russia threatened China, why did not the United States, as in Europe, initiate a new assistance program, comparable to the Marshall Plan, to contain the threat? Truman officials, they charged, were failing to fulfill the logic of their own containment doctrine. The administration replied accurately that it *was* devoted to containment, but it had sufficient reasons for treating Europe and China differently. Some of these explanations were stated publicly, some not.

Never were they pulled together in a comprehensive statement, a failing that brought upon the administration a continued hail of criticism.

Shortly after the launching of the Truman Doctrine and the Marshall Plan, American officials said that the China problem had not reached emergency status; China was not approaching collapse as Greece was. In the spring of 1947 that assessment seemed correct because Nationalist troops were enjoying some success. But after the Communists unleashed their summer offensive and began to drive Jiang's forces back, this argument had to be abandoned. Several other explanations followed. At one time or another from 1947 to 1949 administration spokesmen explained the uniqueness of China by stating that there were far more rebels and Communist sympathizers in China than in Greece; that the Chinese Communists showed some willingness to negotiate (witness the Marshall Mission), unlike the National Liberation Front in Greece; that equipment and goods shipped to China might very likely fall into Communist hands, as was happening; and that a military route to a stable, non-Communist government and peace in Greece seemed feasible, whereas in China it had already proven impossible after two decades of internecine warfare.

Four other reasons were accorded more emphasis. The administration adhered to a Europe-first priority in the Cold War; China was too large, too unmanageable, and had too uncertain a terrain for large-scale American intervention; Jiang Jieshi's regime was resistant to American advice and so corrupt that it squandered American aid; and China did not satisfy an important criterion for receiving a major aid program before the Korean War: A recipient government must have the support of a large percentage of the indigenous population and must be committed to the principle of self-help.

China was not treated as Greece or Europe were because United States Cold War strategy gave first preference to Europe, where the Soviet threat seemed most pronounced. A few days before the announcement of the Truman Doctrine and the proposal to aid Greece and Turkey, Acting Secretary Acheson asked the State-War-Navy Coordinating Committee to study the possibility of aid requests from other governments and to recommend which countries important to American security were neediest. The study seemed necessary because, as his staff informed him, countries other than Greece and Turkey "have cast longing eyes at this new U.S. policy, pointing out parallels they consider might justify assistance in their own cases."[31] The study group's report used the language of the Truman Doctrine and projected American interests as global. Europe was given the highest priority in a list of troubled and needy countries considered of immediate importance to American security. China did not have a place on the list because the study committee could not agree on the question of aid for China. The State Department representative saw no urgency, whereas the War and Navy Department conferees predicted that the State Department's wait-and-see policy might face the United States in the near future with a crisis similar to that in Greece. The JCS named China to a list of countries important to American national security, but they ranked it near the bottom, below Spain and Japan. The JCS concluded that China would be "valuable" in a war "only if we diverted to her great quantities of food and equipment manufactured in this country."[32] The Central Intelligence Agency, on the other hand, apparently concurred with the Europe-first emphasis. In the fall 1948, the agency prepared a long-range forecast indicating that Europe and the Near East would continue as the areas of primary American security interests. This accent on Europe accorded with administration statements that the United States did not have unlimited resources and thus had to be cautious about extending assistance to China so that the more important European Recovery Program was not drained of precious funds.

In addition to the emphasis on Europe as a major reason for administration distinctions between Europe and China was the magnitude of the Chinese crisis and the tasks required to calm it. Senator Henry Cabot Lodge of Massachusetts expressed a common feeling when he said that China was "so damned big."[33] The Director of the Office of Far Eastern Affairs, W. Walton Butterworth, thought Greece a "tea cup" compared to the "oceans" of China.[34] Secretary Marshall shared this uneasiness about the uncertain outcome of American intervention in such a spacious land. Whereas Greece seemed to be a manageable venture, the secretary worried that China would absorb American aid to an unpredictable extent. To eliminate the Communist threat in China, he said, it would be necessary for the United States virtually to take over the Chinese government and administer its economic, military, and governmental affairs—yet that would prove quite difficult because of "strong Chinese sensibilities regarding infringement of China's sovereignty, the intense feeling of nationalism among the Chinese, and the unavailability of qualified American personnel in large numbers required. . . ." In brief, "it would be impossible to estimate the final cost of a course of action of this magnitude."[35] Marshall might also have mentioned that in Greece the United States was stepping into British boots, with already existing institutions through which to exert its influence, and a tradition of tolerated, sometimes welcomed, intervention from the outside. The administration, alert to latent, sometimes explosive Anglophobia in the United States, chose not to use that explanation in selling its Greek program or in urging caution in China.

A third major explanation for the differing treatment of Europe and China was the character and actions of Jiang Jieshi. Unlike comparatively compliant Greek leaders and Europeans eager for American reconstruction dollars, the Generalissimo seemed downright unwilling to accept advice and was insulting in his attempts to deflect American criticism of his regime. "I talked until I became almost embarrassed," Marshall noted a year after the failure of his mission, "and I never got [the Nationalists] to take one step."[36] American officials held little respect for Jiang. He had not fought hard against the Japanese during World War II, instead reserving his army's strength to attack his domestic enemies. He refused to initiate needed, politically popular land reform, and he was unable or unwilling to curb rampant inflation and a dishonest tax system. In addition to wasting large amounts of American money and supplies in a quagmire of corruption, Jiang rejected negotiation or compromise with Mao. He and Madame Jiang continually meddled in American politics and decision-making, deftly cultivating the China Lobby and propagandizing for aid, usually timing their requests for help to moments when the administration seemed least able politically to turn them down, as during the Marshall Plan debates. Jiang ignored strong American advice to reorganize his government on a more representative basis. He sought military assistance yet rejected military advice. The recalcitrant dictator asked for more American money, but, complained American leaders, proved incapable of utilizing it effectively.

The Jiang regime, Marshall concluded, was a reactionary clique, much like "corroded machinery that does not function."[37] Truman privately scored Jiang and his cohorts as grafters and crooks. At the cabinet meeting of March 7, 1947, when aid to Greece and Turkey was discussed just before the Truman Doctrine speech, comparisons with China were outlined. Acheson said the "fundamentals of [the] problems [are] the same. The incidences are different." Both Greece and China were beset by civil war constituting serious Communist challenges to American-backed governments but the environments for meeting the challenges were different. The President then cut in, snapping that "Chiang Kai-shek will not fight it out. . . . It would be pouring sand in a rat hole under [the] present situation."[38] Administration supporter Senator Tom Connally of Texas agreed: "I am at a loss to know what effective measures we could have taken. We gave them credit, we gave them aid. But we could not give them leaders."[39] Near the end of his life, Acheson underscored a point he had made in the White Paper twenty years earlier: "Of course you can't lose something if you didn't have it, and we never had it," he remarked on television in 1969. "China lost itself."[40]

The United States, Acheson pointed out in a diplomatic telegram, had to make a choice of locales and methods for implementing containment. "U.S. cooperation and assistance on even smallest scale must be premised on concrete efforts of country concerned to stand on own feet."[41] In August 1949 he explained to the House Foreign Affairs Committee in executive session that "we can help people who competently and vigorously and faithfully want to help themselves."[42] Kennan echoed this reasoning in a direct comparison between Europe and China. Europeans had a natural will and a strong enough attachment to national independence to resist Soviet pressure.

These prerequisites made it possible for the United States "to follow a program in Europe which proved, I think, much more successful and which looked much more purposeful, much more well-designed probably than what we have done in Asia. . . . "[43] On January 12, 1950, shortly after the Nationalists were driven from the Chinese mainland, Acheson reiterated the self-help theme in an important address to the National Press Club: "American assistance can be effective when it is the missing component in a situation which might otherwise be solved. The United States cannot furnish determination, it cannot furnish the will, and it cannot furnish the loyalty of a people to its government."[44]

For these reasons, then, the Truman Administration distinguished between China and Europe in applying containment and refused to undertake a massive new aid program for China or to dispatch American troops there to halt a Communist victory. By the fall of 1947 it was evident to American officials that Jiang was slipping badly, that a reconciliation of Nationalists and Communists was impossible, and that infusions of aid would be unlikely to reverse the tide. American hesitancy to invest further in Jiang's deteriorating regime surfaced about the same time that Washington was preparing the European Recovery Program. Some members of the Republican eightieth Congress of 1947–48, always alert for opportunities to besmirch the record of the President and unable to exploit the well-received Truman Doctrine and Marshall Plan programs, found the China question a useful stick for beating the administration. Bipartisanship, Vandenberg asserted, did not apply to China. Truman's suppression of the secret Wedemeyer Report of September 1947 further fed criticism that something was amiss in China policy. General Albert C. Wedemeyer had urged extensive American military assistance for China, and Truman's refusal to make his report public aroused suspicion of a cover-up of information damaging to the White House. And the administration introduced in February 1948, with tempered enthusiasm, a China aid bill perhaps intended to quiet critics who might delay action on the Marshall Plan until more assistance was extended to China. Marshall's statement that the new request for $570 million for China was designed to give China only a breathing space did not comfort those who wanted to "save" China. Nor did the sluggish administration of the new assistance. Dewey's defeat in the 1948 presidential election, which seemed to destroy Republican hopes for a change in China policy, further emboldened Truman's political foes to attack the seeming neglect of China.

The critics kept pressing the point that the Truman Administration was not pursuing its own foreign policy of containment. During the 1947 House hearings on aid to Greece and Turkey, congressmen questioned an evasive Acheson about the meaning of the Truman Doctrine with regard to China. He denied that there was a difference between American policy for Greece and for China: "I would like to reiterate," he said, "that I have not said at any time that there is a difference in policy. I was trying to straighten out the actual conditions which I thought were different in the two countries." Acheson affirmed the global impact of containment but begged off from giving specifics about its differing implementation. Representative Judd seized the moment and replied that the Truman Administration was not satisfying the logic of containment: "I do not think we can have one kind of policy in Europe with respect to the danger of Communist-dominated governments and another policy in Asia. . . . "[45]

James Reston of the *New York Times* wrote in 1950 that the Truman Administration appeared inconsistent in "blocking communism in Europe and letting it run wild in Asia."[46] Walter Lippmann, too, thought that Truman officials had brought criticism upon themselves because they had gone to the Congress and the country stating repeatedly that the Truman Doctrine and other measures were containing Communism. "Then the public realized that communism had not been contained in China."[47]

Why was the administration unable to explain in a frank, persuasive manner that it had poured millions of dollars into Jiang's treasury, had armed Jiang's legions to halt the feared Communist victory, had done everything possible to apply containment in China—in short, that the critics were distorting the record? When the administration finally put together its case for not standing in the last Chinese ditch with Jiang, the explanations came too late in the bulky White Paper read by few Americans. It also smacked too much of an apology for failure rather than an explanation for treating Europe and China differently.

Until 1949 Truman officials suffered under the self-imposed restraint that they would not publicly highlight Jiang's ineptitude. Thus, they chose not to explain one of the primary reasons for the difference in applying containment in Europe and Asia. They believed that such a public statement might weaken the beleaguered Jiang regime even further and insure its early collapse. Marshall starkly identified the problem for the Cabinet in November 1948:

> The Nationalist Government of China is on its way out and there is nothing we can do to save it. We are faced with the question of clarifying [this for] the American people and by so doing deliver the knock out blow to the National Government in China—or we can play along with the existing government and keep facts from the American people and thereby be accused later of playing into the hands of the Communists.[48]

Acheson further noted that "you cannot explain why you cannot do anymore, because if you do, you complete the destruction of the very fellows you are trying to help."[49] When Truman released the White Paper, he admitted publicly that certain facts had not been revealed earlier because they "might have served to hasten the events in China which have now occurred."[50] "Our case," regretted the Director of the Policy Planning Staff, "has never been adequately and forcefully presented to the public, largely out of deference to the generalissimo."[51] American leaders believed they had to risk political wounds at home to avoid inflicting them on their collapsing, unregenerate ally Jiang. In so doing, the administration invited charges that it was Washington's fault that the Communists were winning in China and that containment had not been applied vigorously enough. The administration thus had to suffer "misrepresentation, distortion, and misunderstanding," complained the President.[52]

Truman himself might have made public before 1949, the facts damaging to Jiang, but he chose not to. Why? In the final analysis he could not abandon Jiang—could not deliver the knockout blow—because Jiang was America's only real instrument for containing Chinese Communism. The debate between Truman and his critics was not, after all, over whether to abandon Jiang; few advocated that. The question was conditional aid or full-scale support. The administration itself decided by fall 1947 that something had to be done to arrest China's economic deterioration and came to view the China Aid Act of 1948, however futile it might prove in the long run, as a chance to remedy some economic ills and thereby keep Jiang in power longer.

A survey of legislative action reveals that Truman's policy of limited, mostly non-military, non-supervised aid was largely supported by the Congress. The China Lobby's bark was bigger than its bite. The House altered the China Aid Act request for $570 million, providing that $150 million of the amount be set aside for military supplies and that American military advisers, like those operating in Greece, be assigned to China to manage the grant. But the Senate and the administration beat back this attempt to involve the United States more deeply in the civil war. The Senate accepted a figure of $125 million for military supplies, and persuaded the House to approve the amount in the conference bill. But the Senate attached a statement, approved by both houses, that absolved the United States from any responsibility for administering the aid. Furthermore, in February 1949, Senator Patrick McCarran of Nevada introduced a bill to provide a $1.5 billion loan for China for military and economic purposes. The administration spoke vigorously against it, and the Senate thereafter took no action. Another test of strength came in September 1949 when the Senate considered Truman's nomination of W. Walton Butterworth as Assistant Secretary of State. Butterworth had been the Director of the State Department's Office of Far Eastern Affairs and a target of critics such as Senators Bridges and Knowland. Yet the Senate confirmed him in his new post by a vote of 49 to 27. Not only did the Congress not impede Truman's policy of limiting assistance and avoiding direct intervention in the Chinese civil war, it appears that a majority of Americans, as measured in opinion polls for 1948, supported the President in thinking that there was little the United States could do to keep the Communists from assuming power in China.

Congressional critics or hostile public opinion, then, did not and could not force Truman to take important actions he opposed; they did not force him to intervene militarily and did not force him to sponsor a huge aid program for China. Instead, the President pursued his own course, refusing to seek accommodation with Mao and sticking with Jiang, who seemed the only means by which the United States could stem the Communist-Soviet threat it identified in China.

Truman said as early as October 1945, "My policy is to support Chiang Kai-shek."[53] Patrick Hurley had been so instructed when he was posted to China, and Marshall was sent on his mission in 1945 with presidential advice that Jiang was "the proper instrument" and "most satisfactory base" for achieving a unified democratic Chinese government.[54] Marshall, in trying to create a coalition government, never planned that the Communists would supercede Jiang. He and Truman had hoped that Chinese liberals or a middle group would come to power, but such moderates never achieved political standing and the administration found it had no choice but to support Jiang against Mao. Marshall, as Secretary of State, admitted in late 1947 that the United States might have to sustain the corrupt Jiang, however distasteful, because there was no alternative. The Policy Planning Staff also concluded that not much could be done except to "sweat it out" with Jiang.[55] Truman lamented to Senator Vandenberg in 1949: "We picked a bad horse."[56]

What is striking is that the United States backed the wrong horse almost to the very end of the Nationalist regime on mainland China, and, after the outbreak of the Korean War a few months later, continued its backing for Jiang on Formosa. Although some American officials hoped that Titoist tendencies would erode a Sino-Soviet alignment, they did "not think it is safe to bank on it," as Phillip Jessup indicated.[57] The Truman Administration never withdrew recognition from the Nationalist government or attempted to reach a *modus vivendi* with Mao; it never approached the People's Republic of China (proclaimed October 1, 1949) to establish trade relations or to discuss reconstruction aid; and it never recognized the new Communist government, although Britain, Norway, India, and other nations soon did, and still others, like Australia, Canada, and France, preferred recognition but bowed to American policy.

Washington also spurned two Chinese Communist overtures to improve relations. One demarche came in May 1949 from Zhou Enlai, a pragmatic, high-ranking official. He urged closer ties, suggested some disaffection with the Soviet Union, and requested American foreign aid. Truman killed the opportunity by directing the State Department "to be careful not to indicate any softening toward the Communists. . . ."[58] Then, in June, Mao invited Ambassador Stuart to visit him in Beijing; Washington instructed him to avoid such high level contact, and he did not go. Truman did not reject accommodation or recognition just because he feared the China Lobby or negative public opinion. The President for years had successfully defied the Lobby, and he would do so again in 1951 when he fired General MacArthur from his Asian command. As for public opinion, as some members of a State Department roundtable on China commented in October 1949, it "can either be ignored or educated to a new view of the China scene."[59] Truman and Acheson never undertook such a re-education of the American people, because neither man himself held a "new view" of China.

In September of 1949 the United States informed Britain that it would not recognize Mao's government, even though it hoped to play for a split between China and the USSR. The British not only recognized the People's Republic of China but also, at odds with the United States, worked for its admission to the United Nations. London wondered how Washington could drive a wedge between Beijing and Moscow without an American presence in China itself. Indeed, the British sensibly reasoned, American hostility to the People's Republic of China would very likely drive the two Communist states closer to one another. After the failure of the Marshall Mission and years of support for Jiang, about the only way the United States might have contained the Soviet Union was to have come to terms with the Chinese Communist Party, kindling the Titoist inclinations some American officials themselves detected. They rejected that option in favor of non-recognition as a tool of containment. President Truman, by the fall of 1950, seemed to have abandoned all thought of Sino-Soviet differences. British Prime Minister Clement Attlee told Truman in a private conversation that "opinions differ on the extent to which [the] Chinese Communists are satellites." He wondered "when is it that you scratch a communist and find a nationalist." The President would have none of such thinking: "They are satellites of Russia and will be satellites so long as the present Peiping regime is in power." Moreover, they were "complete satellites." And, "the only way to meet communism is to eliminate it."[60]

The Truman Administration practiced containment in China less successfully than in Europe because of peculiar local conditions created by the unreliable Jiang, not because it thought containment was inapplicable to Asia. Truman officials did attempt to use Jiang, through extensive aid, to thwart Mao's ascendency, for in his success they perceived a Soviet threat to China that had global implications. Truman and his advisers invited much of the heavy criticism they received by choosing to sustain Jiang while suppressing official American criticism of him. In their reading of the strategic-economic setback handed the United States by Mao's triumph and by the Sino-Soviet Treaty struck in early 1950, Truman officials opted for a policy of isolation rather than intercourse. After American and Chinese soldiers shed each other's blood in the Korean War, that policy became fixed. America's Asian diplomacy then devoted itself to salvaging Indochina for the French, constructing a regional alliance (Southeast Asia Treaty Organization), reconstructing Japan, and defending Jiang on Formosa. Not until the Nixon presidency, at a time when the Sino-Soviet schism had become cavernous, did the United States act upon a "new view" of China. In 1972 the United States and China began a diplomatic process that culminated in 1979 in a formal exchange of ambassadors. Beijing and Washington took the momentous step toward one another for the same reason: to counter the Soviet threat.

# 5

## Shaping the Cold War Mentality:
## Truman, Public Opinion,
## and Congress

"The President's job is to *lead* public opinion, not to be a blind follower," stated George Elsey, one of Harry S Truman's chief advisers and speechwriters. "You can't sit around and wait for public opinion to tell you what to do." He grew more emphatic: "In the first place, there isn't any public opinion. The public doesn't know anything about it; they haven't heard about it. You must decide what you're going to do and do it, and attempt to educate the public to the reasons for your action."[1] This unabashed stress on the President's initiative in foreign policy and the self-conscious notion that the "public" must be coaxed or "educated" into supporting what the chief executive has already decided, underscores the point that the President enjoyed considerable freedom in the making of foreign policy. One student of the topic has graphically described the President as a "kind of magnificent lion who can roam widely and do great deeds so long as he does not try to break loose from his broad reservation." The "restraints" of his domain are "designed to keep him from going out of bounds, not to paralyze him in the field that has been reserved for his use."[2] Few Presidents, springing as they do themselves from the American mind and spirit, have strayed "out of bounds," and few have ever been paralyzed in their efforts to achieve their self-defined "great deeds." Harry S Truman was no exception when he attempted to meet the perceived Soviet threat and to arouse public support for his containment policies.

In the early years of the Cold War, public opinion and the Congress set very broad and imprecise limits on presidential activity in international affairs. Foreign policy initiative lay with the executive branch. The administration's diplomacy was not determined by the buffeting winds of public sentiment or by an obstructionist Congress. Seldom did Truman have to do what he did not want to do; seldom did he have serious trouble over-coming the obstacles of a parsimonious Congress or interest groups in order to pilot his foreign programs through either a Democrat- or Republican-controlled Capitol Hill. Congress was generally compliant, and the American people were "yea-sayers." Although Truman officials occasionally suggested that they were sensitive to or influenced by public attitudes, the minutes and records of such high-level, policy-making groups as the cabinet, the Committee of Three (Secretaries of State, War, and Navy), and the Secretary of State's Staff Committee, as well as official diplomatic correspondence, do not reveal that American leaders paid much attention to American public opinion or that they were swayed by it to do something they did not wish to do. President Truman charted his own foreign policy course and successfully persuaded the reluctant to walk his path.

Some scholars have attributed more importance to public opinion than Truman did. On the other hand, others have been skeptical about the public's power over leaders. Bernard C. Cohen, for one, has written that "even though foreign policy leaders have widely underestimated their freedom of maneuver in foreign policy, they still perceive that freedom more accu-rately than many scholars have, and more accurately even than they themselves usually admit openly."[3] Clinton Rossiter has marveled at Truman's "spacious understanding" of presidential power, despite the President's often humble, folksy explanation of his role.[4] Truman once wrote, for example, that "all the President is, is a glorified public relations man who spends his time flattering, kissing and kicking people to get them to do what they are supposed to do anyway."[5] This self-deprecating and disingenuous statement did not accord with the facts. In fact, Truman was a maestro in drawing full tones from his constitutional strings and from his public chorus.

Certainly Truman and his advisers worried about public opinion. They read the polls. They courted the ethnic vote. They buttonholed congressmen. They warned against the dangers that could come from negative opinion. In 1947 Clark Clifford urged his boss to name the new European Recovery Program the "Truman Plan." "Are you crazy," Truman interjected. "If we sent it up to that Republican Congress with my name on it, they'd tear it apart. We're going to call it the Marshall Plan."[6] Thus, Truman officials were sensitive to the public pulse, but they believed that public opinion and Congress could be persuaded and cultivated on most issues—that they were permissive rather than restrictive. The President's power to create and to lead public opinion was studiously exercised. Any President can pre-empt the airwaves to make his case. Reporters flocked to Truman's press conferences, and as James Reston of the *New York Times* witnessed, "when the press conference ends, the scramble of reporters for the telephones is a menace to life and limb. . . . "[7] The President grabbed headlines; even his daily walks, colds, and piano playing became news.

The President exploited frightening world events in order to garner support for his foreign policy. Truman once remarked that without Moscow's "crazy" actions, "we never would have had our foreign policy . . . we never could have got a thing from Congress."[8] He exaggerated, but what he meant is illustrative: He exploited Cold War tensions through a frequently alarmist, hyperbolic, anti-Communist rhetoric that he thought necessary to insure favorable legislative votes, to disarm his critics, and to nudge budget-conscious congressmen to appropriate funds for his programs.

American leaders complained during and after the war about lingering American "isolationism" and bemoaned the work needed to convert it to "internationalism." During the war, commentators on the American mood predicted that at war's end the GI would exchange his fatigues for "isolationist" garb. Reporters for *Time* magazine who asked soldiers what they thought about the war received the answer that they "never wanted to hear of a foreign country again."[9] A serviceman's indelicate ditty went this way:

> I'm tired of these Limeys and Frogs,
> I'm fed to the teeth with these Gooks, Wops, and Wogs.
> I want to get back to my chickens and hogs,
> I don't want to leave home any more.[10]

After the war, W. Averell Harriman remarked that many Americans wanted nothing more than to "go to the movies and drink Coke," and Dean Acheson, disliking the United States' rapid postwar demobilization, defined the administration's task in 1946 as "focusing the will of 140,000,000 people on problems beyond our shores. . . . "[11] Senator Wayne Morse of Oregon, a firm supporter of Truman's "hard-boiled policy" in 1945, feared that "sentimentalist groups" would once again gain control of American foreign policy.[12] "Americans can no longer sit smugly behind a mental Maginot line," implored Truman just before the end of the war.[13]

"Isolationism" actually evaporated quickly. "Isolationism"— that vague feeling that the United States should restrict its activities overseas, especially in Europe, follow an independent or unilateral course, and mind its own business—collided with the realities of the air age and an international system which cast the United States in a starring role. The Pearl Harbor attack, said Republican Senator Arthur Vandenberg of Michigan, "ended isolationism for any realist,"[14] As he put it in early 1945, when he renounced his own "isolationism" on the Senate floor, "I do not believe that any nation hereafter can immunize itself by its own exclusive action. . . . Our oceans have ceased to be moats which automatically protect our ramparts."[15] Not wishing to repeat the post-World War I debacle wherein President Woodrow Wilson launched the League of Nations and the United States then jilted it, Americans stood with Vandenberg in repudiating a failed past. Leaders and publicists of many political persuasions lectured the American people that the "horse-and-buggy days are gone," for "in a world in which a man can travel from New York to India in less time than it took Benjamin Franklin to travel from Philadelphia to New York, the attempt to escape into the Golden Age of normalcy is an invitation to chaos."[16] Such lectures, joined with the fast pace of global crises and presidential statements about the tremendous American responsibility in a broken world, undercut or crippled isolationist feeling. American membership in the new United Nations Organization was roundly approved. Public opinion polls recorded that large majorities of Americans applauded an activist foreign policy abroad. Even some of the so-called postwar isolationists, like Senator Robert Taft of Ohio, who tried to curb American programs designed to build an economic and military shield around Western Europe, seemed inconsistent as they shouted for American interventionism in Asia. As the pacifist A. J. Muste once remarked, "For isolationists these Americans do certainly get around."[17] Thus, although Truman and his officers worried about isolationism, they did not find it an obstacle to their policy of "getting tough" with the Soviet threat.

It was not "isolationism" which characterized American public opinion about Cold War events but ignorance or indifference. "Public opinion" suggests in a vague way that "the people" express themselves collectively on issues of national importance; yet on topics of foreign policy there was no mass public that spoke out on international relations with force of unity or that could seriously instruct or influence leaders. Most Americans read little about foreign events in their newspapers, which devoted the greater proportion of their columns to domestic topics; consequently, they were ill-informed about politics abroad. When asked to list the most significant issues before the nation, Americans catalogued domestic problems. In the Truman years, labor strikes, reconversion, price controls, housing shortages, and inflation occupied Americans. "Foreign affairs," grumbled a blue-collar worker. "That's for people who don't have to work for a living."[18]

This striking ignorance—sometimes mistaken for isolationist thought—was demonstrated in a study that the Council on Foreign Relations commissioned in March of 1947. The researchers for *Public Opinion and Foreign Policy* (1949) found to their dismay that three out of every ten American voters were unaware of almost every event in United States foreign relations. They concluded further that only one-quarter of the American electorate was reasonably well informed. Sixty-five out of every one hundred voters admitted that they rarely discussed foreign affairs. Other studies reported that in 1946, 43 percent of adult Americans had not followed the discussion on an American loan to Britain, that 31 percent could not even give a simple answer to a question about the purpose of the United Nations Organization, and that 58 percent had not paid attention to the major debate between the Truman Administration and Henry A. Wallace which led to the latter's explosive departure from the cabinet. In a Gallup survey for September 1949 some 64 percent of those polled had neither heard nor read anything about the controversial China "White Paper." Students of the relationship between foreign policy and public opinion have found these figures consistent with American tradition. "The mass public," one has concluded, "is uninformed about either specific foreign policy issues or foreign affairs in general. Its members pay little, if any, attention to day-to-day developments in world politics."[19] This ignorance or apathy permitted Washington officials wide latitude and independence in making foreign policy to meet the Communist threat. Elsey put it too strongly when he said, "there isn't any public opinion." He would have been more accurate had he said that public opinion was so weak that leaders were neither educated nor moved much by it.

There was a "public opinion" that Truman and his advisers took seriously, and diligently sought to cultivate—the opinion of the approximately 25 percent of the American people who were attentive to foreign policy questions. Scholars have called them variously "notables," "opinion leaders," or the "foreign policy public." They constituted the small number of Americans who studied the foreign news, who traveled abroad, who spoke out. They held positions in American society that commanded authority and insured influence—journalists, businessmen, labor leaders, intellectuals, and members of various interest and citizen groups. They produced the "public opinion" on foreign policy issues that counted. A Truman officer summarized the point: "It doesn't make too much difference to the general public what the details of a program are. What counts is how the plan is viewed by the leaders of the community and the nation."[20] Another State Department official remarked that "we read the digests, we ponder the polls, and then we are likely to be influenced by our favorite columnist."[21] The "foreign policy public" was important, too, because it could influence a wider audience. As someone put it, he "who mobilizes the elite, mobilizes the public."[22] The Truman Administration happily found that the elite endorsed the President's foreign policy, further enhancing his freedom in policymaking.

To cement the alliance between the "opinion leaders" and the administration, government officials wooed them. Interest group executives were given flattering appointments on consulting bodies like the President's Committee of Foreign Aid, the Public Advisory Board of the Economic Cooperation Administration, and the Business Advisory Council of the Department of Commerce. They served as consultants to the American delegation at the founding meeting of the United Nations Organization in San Francisco. They were appointed to high offices, as when Paul Hoffman of Studebaker Corporation was named to head the Economic Cooperation Administration. They were invited to appear before congressional committees to state their views as "experts." They enjoyed special State Department briefings or attended special White House conferences like that for the Marshall Plan on October 27, 1947. Some individuals were consulted so often or sat on so many public boards that they could truly be called "external bureaucrats."[23] All in all, they were courted, sometimes because their views provided new insight, but largely because they represented a "public opinion" that mattered, that could be shaped to endorse Truman's Cold War policies.

The major task of cultivating the elite fell to the State Department's Office of Public Affairs, which was organized near the end of World War II. This agency was designed to strengthen relations with public groups and prominent individuals. It maintained ties with over two hundred organizations and drafted State Department responses to letters from opinion leaders. So that it could anticipate criticisms, it conducted polls and interviews with American voters and subscribed to over one hundred newspapers and magazines. The office also sponsored an annual conference in the Department of State to which it invited about two hundred people—usually presidents of national associations. Smaller meetings of ten to thirty participants were held each week: Usually the conferees met with middle-echelon officers, but on one occasion Secretary Acheson himself addressed a small group to explain his controversial policy toward China. The Director of the Office of Public Affairs, Francis Russell, frankly recalled that the thin line between propaganda and education was sometimes crossed. Whether the office's function was education or propaganda, the chief purpose was evident: to sell the President's foreign policy to opinion leaders. That purpose was met.

A sterling example of cooperation between the government and the elite came in 1947–48, when the Committee for the Marshall Plan was organized to arouse public support for the European Recovery Program. Former Secretaries of War Henry L. Stimson and Robert Patterson joined former Assistant Secretary of State Dean Acheson to launch the committee in the fall of 1947. Working closely with the State Department, the committee staff ran newspaper ads, circulated petitions, organized letter campaigns to congressmen, and maintained an active speaker's bureau. Groups like the Farmer's Union that were asked to testify before congressional committees were supplied with prepared texts. The presidential assistant Richard E. Neustadt lauded the committee as "one of the most effective instruments for public information seen since the Second World War. . . . "[24]

The President himself attempted to shape the public opinion that his administration wanted to hear. He simplified issues, stirring the heart more than the mind. His techniques have been common to most presidencies: He spoke at times in alarmist terms, predicting dire results if a certain policy was not carried out; he appealed to patriotism to rally Americans to his banner, often recalling the sacrifices of World War II; he created awesome and frightening images of the foreign adversary. He also sketched pictures of political enemies like Henry A. Wallace which suggested that "Reds" might someday roam the corridors of the White House. Truman and his advisers were unrelenting toward Wallace, practicing a highly emotional "Red-baiting." To one congressman, the President wrote in 1947 that Wallace "seems to have obtained his ideas of loyalty . . . from his friends in Moscow and, of course, they have no definition for that word."[25] Several months later Clark Clifford planned strategy for the 1948 presidential campaign. His advice reflected his mentor's preference: "Every effort must be made *now* jointly and at one and the same time—although, of course, by different groups—to dissuade him [Wallace] and also to identify him in the public mind with the Communists."[26]

The historian Thomas A. Bailey, author of a popular 1948 book entitled *The Man in the Street*, defended the President's efforts to "educate" an ignorant public: "Because the masses are notoriously short-sighted, and generally cannot see danger until it as at their throats, our statesman are forced to deceive them into an awareness of their own long-term interests. . . ." He went on: "Deception of the people may in fact become increasingly necessary, unless we are willing to give our leaders in Washington a freer hand. . . . The yielding of some of our democratic control over foreign affairs is the price that we may have to pay for greater physical security."[27] Charles Bohlen noted that Cold War national security required "a confidence in the Executive where you give human nature a very large blank check."[28]

Because the American people usually rallied around their leaders during foreign policy crises, and because the President usually succeeded in cultivating or manipulating public sentiment, it is not surprising that the public opinion the White House and the State Department heard largely matched the public opinion it worked to create. The administration in essence listened to the echo of its own words. "What the government hears," suggested one official, borrowing a metaphor from the Navy, "is really the sound of its own screws, reflecting off its own rudder and coming up through its own highly selective sonar."[29]

Still, administration figures said that public opinion counted, that it guided their foreign policy. They were saying so ritualistically, out of habit and necessity, not because it was reality. One would expect high-level officials in a democratic-representative political system to say publicly that they believed the public's views counted: that is what a public audience wanted to hear. Telling the public so actually constituted another component in the effort to create friendly opinion.

American diplomats sometimes told foreign leaders that the United States could not undertake a certain policy or program because public opinion or the American people would not countenance it. In his famous acrimonious exchange with Molotov in April 1945, Truman warned that Russian behavior would affect United States policy on foreign aid to the Soviet Union, because "he could not hope to get these measures through Congress unless there was public support for them."[30] Acting Secretary of State Joseph Grew told the Yugoslav foreign minister a month later that in matters of foreign policy "we were guided by public opinion" and that Yugoslavia's request for foreign aid "would to a large measure depend on the impression which the American public will gain from the policies and events in the countries recently liberated."[31] In 1947 Under Secretary of State Will Clayton said much the same thing in his attempt to persuade the British to join a European economic plan. In all three examples, however, there is insufficient evidence to argue that the administration's reading of public opinion influenced its foreign economic policies. Such statements, it appears, more often than not represented a diplomatic device designed to press foreign officials rather than a political reality at home.

Public opinion usually supported Truman's early Cold War policies, giving him a free hand. Even hostile opinion did not deter him from doing what he thought had to be done. Good examples are the questions of loans to Russia and Britain in the fall of 1945. Sixty percent of the respondents in a Gallup poll disapproved of a loan to Russia (only 27 percent approved, and 13 percent had no opinion). Another Gallup poll revealed exactly the same statistics of public opposition to a loan to Britain. The Truman Administration nevertheless proceeded, in apparent violation of "public opinion," to negotiate a $3.5 billion loan to Britain, while neglecting negotiations with the Russians, who received no loan. In April of 1947, shortly after his Truman Doctrine speech, the President found that a large majority of Americans felt that the problem of aid to Greece and Turkey should be turned over to the United Nations. He never followed such a policy, again in apparent defiance of "public opinion." What he and the State Department did instead was to launch an effective public campaign with the message that the United Nations was an infant organization not yet ready to take on the momentous task proposed. To quiet discontent in Congress, the administration endorsed an innocuous amendment drafted by bipartisan leader Senator Arthur H. Vandenberg and the State Department staff to the effect that the United Nations could assume the task when it was ready. "I never paid any attention to the polls myself," remembered Truman, "because in my judgment they did not represent a true cross section of American opinion. . . . I also know that the polls did not represent the facts but mere speculation. . . ."[32]

When the administration said that it was responding to public opinion, it usually meant it had to deal with a special-interest group that was one-issue oriented and like the President, was often able to exploit indifference to satisfy its aims. Thus, in one of the rare measurable examples of successful interest-group influence, a well organized and vocal Jewish-American community exerted considerable pressure on vote-conscious Truman in the 1948 election, and he reacted by over-ruling his own State Department and backing the establishment of the new state of Israel.

Even highly-charged special interest groups, however, did not always succeed in influencing foreign policy, as the case of the Polish-Americans and United States policy toward Poland illustrates. At the Teheran Conference of 1943, President Roosevelt told Premier Stalin that he could not take part in any public agreement over Polish boundaries, because there were six to seven million Polish-Americans in the United States and that "as a practical man, he did not wish to lose their vote."[33] In June of 1944, Breckinridge Long, a State Department official and political appointee of Roosevelt, worried about the Soviet presence in Poland and recorded in his diary that the Poles held the political balance in Illinois and Ohio and were politically potent in Detroit, Chicago, and Buffalo. In July of the following year, President Truman mentioned at the Potsdam Conference that there were six million Poles in America who could be dealt with much more easily if a free election was held in Poland.

Although American diplomats were sensitive to this conspicuous political reality, American policy toward Poland was not shaped or determined by Polish-Americans. Several reasons explain this. The figure of six to seven million Polish-Americans was inflated, for it included Poles of all ages; their voting strength was much less significant. Few Poles were in positions of national leadership from which to influence policy: No Americans of Polish descent sat in the Senate until well after World War II—although Michigan Senator Vandenberg spoke for them—and there were only ten to twelve Polish-Americans in the House. Also, political leadership in the Polish-American community itself was splintered, weakening its impact. Neither President respected the Polish-American Congress as an interest group, and they were able to reduce the effect of Polish-American opinion. Roosevelt deftly courted Polish-American leaders in 1944 by speaking in generalities and by apparently convincing them that he agreed with them. In the election of 1944, despite differences with Roosevelt, the Poles stuck with the Democrats in overwhelming numbers. Truman grew angry with criticism from Polish-American leaders like Charles Rozmarek that the United States had sold out Poland to the Soviets, and he paid less and less attention to them. Although Rozmarek and other leaders defected in 1948 from the Democratic ranks, Polish-Americans, as ardent Democrats, on the whole voted for Truman. In short, although the Presidents and their advisers naturally worried about and courted Polish-American votes, the Polish-American community did not wield important influence in the shaping of American foreign policy. Truman pursued *his* policy toward Poland.

If public opinion did not have much of an impact overall on Truman's foreign policy, can it be argued that Congress did? Congress obviously possessed powers that the amorphous public opinion lacked. In a negative and abstract sense, Congress had the power to defeat a presidential proposal. The Senate could reject a treaty if one-third of its members (plus one) voted "nay," and the House of Representatives, gripping the purse strings, could refuse to appropriate funds. Both houses of Congress had the power to investigate. Administration officers had to troop into congressional hearings to answer questions from suspicious legislators, some of whom harbored stereotypes of "striped pants" diplomats. For a six-month period in 1947, the State Department expended more than one thousand man-days in describing and defending Truman's policies before congressional committees, and when he was Secretary of State, Acheson spent about one-sixth of his working days in Washington meeting with congressmen.

Congress can also create "watchdog" committees to oversee the execution of programs that it approved; the Senate votes on presidential appointees to diplomatic posts; the Constitution empowers the Congress to set tariffs, regulate foreign commerce and immigration, and declare war; and, in extreme cases, the legislative branch can exercise its impeachment powers. In short, the "separation of powers" invests Congress with wide potential authority in making and conducting foreign policy, and any administration—if it wishes to push its programs through the legislative process—must be alert to congressional opinion. In practice, however, Congress has been largely subservient to the President in matters of foreign policy.

Most Presidents have reached into well-stocked arsenals for devices to augment their powers vis-à-vis Congress. Truman was a former senator, with ten years in rank, and he utilized his status as an alumnus to gain access to the upper house. He called upon his friends there, strengthening personal contacts. As President he could be found having lunch in the Senate dining room. Once, on July 23, 1947, he entered the Senate chamber itself, prompting presiding officer Vandenberg to announce that the "ex-Senator from Missouri is recognized for five minutes."[34] Besides personal lobbying, Truman attempted to persuade the larger electorate, which would, it was anticipated, in turn exercise some influence over congressmen worried about what the folks back home were feeling. The administration could further shape favorable congressional opinion by including congressmen directly in diplomatic negotiations. Thus, Senators Robert Wagner and Charles Tobey were present at the Bretton Woods Conference, Vandenberg joined the American delegation to the United Nations Conference in San Francisco, and Vandenberg and Senator Tom Connally served at the 1946 Paris Peace Conference. Congressmen were carried by military aircraft to economically hobbled postwar Europe to encourage "aye" votes for Truman-initiated recovery programs. In June of 1949, just before the Senate approved American membership in NATO, Secretary of Defense Louis Johnson told the cabinet that a "liberal attitude in ferrying members of Congress to Europe" would be "helpful in developing support for [the] Atlantic Pact and Arms Program."[35]

Because the conduct of foreign policy has always operated in clouds of secrecy, any administration can withhold information from Congress on grounds of national security or executive privilege. "There is a point . . . when the Executive must decline to supply Congress with information," Truman wrote, "and that is when he feels the Congress encroaches upon the Executive prerogatives."[36] Acheson believed that Congress was "dependent" upon the executive for knowledge of events—the "flow of papers"—and that "here knowledge is indeed power."[37]

While Truman was in office, initiative in foreign policy lay with the President. In essence, he decided which diplomatic topics the Congress acted upon. He sent requests for foreign aid; he delivered special messages; he outlined new programs. In short, as the famous saying has it, "the President proposes, the Congress disposes." What that often meant in the Truman period was that Congress gave legitimacy to what Truman had already decided. By announcing policies and then going to Congress to ask for endorsement, the President handed the legislative branch *faits accomplis*. By gaining the initiative in this way, he could exert considerable pressure on congressmen who did not wish to be placed in the unenviable position of "naysayers" or "neo-isolationists"—especially on Cold War issues which he defined in exaggerated terms as matters of national survival.

The President's March 1947 request for aid to Greece and Turkey is a case in point. Once Truman had enunciated his "doctrine" before a joint session of Congress, many of its members hesitated to deny him his program. Senator Owen Brewster of Maine was reluctant "to pull the rug from under his feet," and Senator Leverett Saltonstall of Massachusetts said that Congress had to support the President, or "many people abroad who do not fully understand our system of government would look upon our failure as a repudiation of the President of the United States. American prestige abroad means more security and safety at home; this is to me a compelling reason."[38] Legislative critics questioned the apparent indiscriminate globalism of the doctrine, the bypassing of the United Nations, the aid to conservative regimes, the salvaging of the British sphere of influence, and the cost, yet sixty-seven senators of the Republican Eightieth Congress stood with the President and only twenty-three opposed him. Republican Senator Henry Cabot Lodge thought that his colleagues had to decide whether or not "we are going to repudiate the President and throw the flag on the ground and stamp on it. . . . "[39] Vandenberg was particularly alert to the relationships among the President's methods, congressional responsibilities, and American foreign policy in the Cold War:

> The trouble is that these "crises" never reach Congress until they have developed to a point where Congressional discretion is pathetically restricted. When things finally reach a point where a President asks us to "declare war" there is nothing left except to "declare war." In the present instance, the overriding fact is that the President has made a long-delayed statement regarding Communism on-the-march which *must* be supported *if* there is any hope of ever impressing Moscow with the necessity of paying any sort of *peaceful* attention to us whatever. If we turned the President down—after his speech to the joint Congressional session—we might as well either resign ourselves to a complete Communist-encirclement and infiltration or else get ready for World War No. Three.[40]

Bipartisanship was another resource for engineering congressional consent for presidential foreign policy. Republicans, joining the President in his extreme depiction of the world crisis and the need to restore a broken world under Communist threat, muted their criticism of Truman's foreign policy, except that toward Asia. As Republican Senator Alexander Smith of New Jersey told Senator Robert Taft of Ohio—one of the few Republicans who vigorously questioned the bipartisan consensus—"the President and Mr. Byrnes have some tough nuts to crack and I feel it is vitally important for them to have a united nation behind them."[41] Senator Vandenberg, one of the architects of bipartisanship, boasted to a Detroit audience in early 1949 that "during the last two years, when the Presidency and Congress represented different parties, America could only speak with unity. . . . So-called bipartisan foreign policy provided the connecting link. It did not apply to everything— for example, not to Palestine or China. But it did apply generally elsewhere." He concluded: "It helped to formulate foreign policy *before* it ever reached the legislative stage."[42] Vandenberg chaired the Foreign Relations Committee during the Eightieth Congress (1947–48) and had something to say on almost every issue. He nursed presidential ambitions, became the foreign policy spokesman for his party, and wanted to be consulted by the Truman Administration. A chronicler of the diplomatic career of Dean Acheson has concluded that Acheson "unfairly considered the Senator a superficial thinker, and egoist. . . . This egotism, Acheson believed, made it possible to manipulate Vandenberg by giving him the illusion of victory on an aspect of an issue."[43]

The Truman Administration deftly flattered Vandenberg, appointing him to delegations, calling upon him for advice, applauding his Cold War patriotism, agreeing to his suggestions for appointments. During the launching of the Marshall Plan, Vandenberg said he would not support the program unless a distinguished panel of citizens was appointed to study it. Truman complied and named W. Averell Harriman to head the committee. Truman wanted to make Acheson the administrator of the European Recovery Program, but Vandenberg disapproved. Truman thereupon consulted with Vandenberg and appointed one of his candidates, Paul Hoffman, president of Studebaker. In both cases, the Marshall Plan itself was protected from serious challenge. The famous "Vandenberg Resolution" of 1948 was first suggested by and written in collaboration with the State Department. This document recommended that the United States associate itself with regional security pacts like the pact just formed in Western Europe under the Brussels Treaty. Writes one scholar, "the adoption of the Vandenberg Resolution [by a Senate vote of 64-4] indicates once again the executive's primacy in the identification and selection of problems which occupy the foreign policy agenda of Congress and the executive."[44] All in all, Truman officials shrewdly and easily cultivated Vandenberg, disarming a potential critic and insuring bipartisanship.

In the election of 1948, bipartisanship further greased political tracks for presidential foreign policies. As in 1944, Republican leaders pulled their punches. John Foster Dulles and Vandenberg persuaded their Republican party colleagues, including candidate Thomas Dewey, to refrain from criticizing the foreign policy of the Democratic President. "One of the things I tried to keep out of the campaign was foreign policy," remembered Truman.[45] He succeeded.

"Bipartisan foreign policy is the ideal for the executive," mused Dean Acheson, "because you cannot run this damned country any other way except by fixing the whole organization so it doesn't work the way it is supposed to work. Now the way to do that is to say politics stops at the seaboard—and anyone who denies that postulate is a son-of-a-bitch and a crook and not a true patriot. Now if people will swallow that, then you're off to the races."[46] Votes in the Senate on key postwar programs reveal that the Truman Administration commanded the results of the "races." The Bretton Woods agreements (World Bank and International Monetary Fund) passed 61-16, United Nations Charter 89-2, ratification of the Italian peace treaty (1947) 79-10, assistance to Greece and Turkey 67-23, Rio Pact 72-1, Interim Aid to Europe 86-3, European Recovery Program 69-17, NATO 82-13, and ERP Extension 70-7. Only the favorable vote on the British loan of 1946 was close—46-33.

Most members of Congress debated how much to spend, not whether to spend. Sometimes they trimmed budgets and forced the executive to compromise on administrative machinery for foreign aid programs. Sometimes the President thought that he had to speak with alarm to persuade Congress to give him the votes. This was especially true in 1947–48 when the Republican Eightieth Congress sat; yet this was the same Congress that met every one of his requests for expensive foreign aid projects. Truman's tangle with what he called the "do-nothing" Eightieth Congress stemmed from domestic issues, not questions of international relations. After Truman's stunning victory in the 1948 election, Mao Zedong's 1949 triumph in China, the emergence of McCarthyism in 1950, and the outbreak of the Korean War in June of that year, bipartisanship eroded. Still, the administration, despite congressional cuts in military assistance, got what it wanted from Congress, including aid to Communist Yugoslavia after Tito's break with Stalin. Acheson, no doubt a wry smile spreading across his mustachioed face, recalled that "many of those who demanded the dismissal of the Secretary of State in 1950–52 joined in passing all the major legislation he laid before Congress. . . ."[47]

Most of the acts and treaties approved by Congress placed immense power in the hands of the President and his subordinates—to allocate funds abroad, to direct military assistance, to react quickly to crises, to order the use of nuclear weapons, to negotiate tariffs, to appoint administrators. Then, too, the President circumvented Congress by making executive agreements. Truman signed eighteen military executive agreements during his tenure—for example, those that permitted the United States to use an airbase in the Azores (1947), to place troops in Guatemala (1947), and to hold bases in the Philippines (1946). In other important cases, Truman simply did not go to Congress to ask for approval of his decisions. He decided to order the dropping of the atomic bombs on Hiroshima and Nagasaki on his own, and he alone possessed the awesome power thereafter to authorize the use of nuclear weapons. The Berlin Blockade was met not with a congressional program but with the President's policy of an airlift. Truman never went to Congress, moreover, for a declaration of war during the Korean War, but Congress voted funds time and time again to continue it. Acheson explained that the administration did not ask Congress for a war declaration because it did not want to invite hearings which might produce that "one more question in cross-examination which destroys you, as a lawyer. We had complete acceptance of the President's policy by everybody on both sides of both houses of Congress." He simply did not want to answer "ponderous questions" that might have "muddled up" Truman's policy.[48]

Public opinion and the Congress proved malleable, compliant, and permissive in the making of America's Cold War foreign policy. More than once President Truman successfully grasped the opportunity to free himself from political and constitutional restraints so that he could define and carry out his foreign policy preferences. Sam Rayburn, Speaker of the House in 1940–47 and again in 1949–53, thought that it had to be that way: "America has either one voice or none, and that voice is the voice of the President—whether everybody agrees with him or not."[49] Indeed, the Soviets listened not to the voice of Congress or to the many voices of American public opinion but to the tough words of President Harry S Truman.

# 6

# Resisting Exaggerations of the Threat: Critics of the Early Cold War

American foreign policies and the wars they helped to initiate or failed to avert have always had their largely unwelcome and unappreciated, yet valuable critics—anti-imperialists in the 1890s, dissenters from American entry into World War I, isolationists in the 1930s, and doves in the 1960s. Their dissent marked them in the popular and even official mind as ignoramuses, deluded citizens, conscious troublemakers, irrational misfits, or dangerous traitors. Thus Senator Robert M. LaFollette, Jr., boomed Theodore Roosevelt in 1917, was "loyally and efficiently serving one country—Germany" by questioning American participation in the First World War.[1] Critics of the Vietnam War were charged by the architects of that conflict with assisting, through their vociferous dissent, the cause of an aggressive North Vietnam. Critics, in asking unpopular questions and proposing alternatives which often differed radically from official policy, refused to accept the consensus that leaders have always sought in times of crisis, and hence became conspicuous targets for contempt and verbal abuse.

During the early years of the Cold War, many American critics resisted Truman's shaping of the Cold War mentality and had to endure this pattern of international crisis, consensus, risky dissent, and intolerance. Nasty name-calling, red-baiting polit-

95

ical campaigns, new internal security measures, and political
ostracism were visited upon them. Moreover, they were
thought to be unrealistic, misguided, wrong-headed, and, in
some cases, even duped and disloyal. A common theme, too,
was that the most ardent dissenters apologized for and excused
the brutalities and offenses of Stalinist foreign policy. This
unfavorable judgment was sustained by the official explanation
that the coming of the Cold War was inevitable; by the power of
bipartisan foreign policy and consensus in the Truman years; by
the apparent life-or-death struggle with Communism which
called forth patriotic support for the administration; and by the
widespread belief in the righteousness of America's activist
world role. The carelessness, contradictions, and strained rhetoric of some of the dissenters themselves also contributed to the
negative portrait.

The Cold War critics were, of course, never organized into one
group; they were quite divergent in backgrounds and attitudes,
and sometimes in vigorous disagreement with one another.
What united most of them was not their desire for peace, for
almost every American wanted that, but their generally shared
belief that reconciliation with the Soviet Union was possible
without sacrificing the national interest, that American foreign
policy and its exaggerations were helping to aggravate Soviet-
American friction, that the Cold War should not be militarized,
and that the conflict threatened American institutions and
cherished principles at home. These critics recognized ideological differences between Russia and the United States, but
believed that some kind of *modus vivendi* or coexistence without
military escalation could be realized diplomatically. Many of
them also denied that Russia was such an aggressive monster
that the United States should, as it seemed to be doing, avoid
negotiations.

Some of the critics sat in the United States Senate. Robert A.
Taft, "Mr. Republican," became a sober dissenter from exaggerated views of the Soviet Union. A strong anti-Communist, he
asked tough-minded questions about the exact nature of the
Soviet threat, demanding evidence in specific cases. Taft was a
fiscal conservative, of course, and that in part explained his
opposition to many Truman policies. But, more, he feared

Truman would provoke a war. Although Taft reluctantly supported the Truman Doctrine, he rejected hyperbolic claims that the Russians were hell bent on military aggression. Senator Taft stoutly opposed the creation of NATO and the positioning of American troops in Europe as unnecessary, because, he argued in his book, *A Foreign Policy for Americans* (1951), the Soviet threat was ideological not military. Taft seldom agreed with his colleagues Claude Pepper of Florida and Glen Taylor of Idaho, for their liberalism challenged his conservatism on domestic issues, and their often shrill language, excited oratorical performances, and acquaintances with blunt dissenters on the left offended the sedate legislator from Ohio. Their contrasts were evident in 1948, for example: Taft supported Thomas Dewey; Pepper, a Democrat, reluctantly backed Truman; and Taylor left the Democratic party to run as the vice-presidential candidate for the Progressive party pledged to Henry A. Wallace. Nonetheless, Taft, Pepper, and Taylor stood together in challenging the expansion of American military power abroad.

The journalists and Cold War critics Walter Lippmann and I. F. Stone could not have been much more different from one another in temperament and style. Lippmann was a highly respected commentator with a regular, syndicated column. He had access to decision-makers whom he cultivated, and his criticisms often amounted to gentle slaps. Yet the critique of the containment doctrine in his book *The Cold War* (1947) still stands as an authoritative, perceptive analysis of globalism and a prophetic warning against over-commitment abroad. Stone was a brazen, insistent, and unorthodox investigative journalist who was never accepted into his profession's elite. Lippmann published his many books with prominent houses like Harper's and Little, Brown; Stone had to scratch around before finding the Monthly Review Press to publish his provocative challenge to the official explanation of the origins of the Korean War as Soviet aggression, *The Hidden History of the Korean War* (1952).

Another critic, Henry A. Wallace, had traveled many political routes from agricultural Iowa to Washington, liked dramatic public attention, and styled himself as one of the apostles of the "common man." Less guarded and reflective than Lippmann or Taft, Wallace had a lively tongue which alienated him from the political establishment and often led his listeners to ask for clarification. Truman fired Wallace from his post as Secretary of Commerce in 1946 for openly criticizing the President's get-tough policy. Wallace then went on to run very poorly as the Progressive party candidate for the presidency in 1948.

A wide spectrum of black critics, with a perspective different from that of other dissenters, centered their attention on colonialism in Africa. There was unity in their complaint that the United States identified too closely with the imperial policies of its white Cold War allies in Europe, but there was disagreement between the radical Council on African Affairs and the more cautious National Association for the Advancement of Colored People. Walter White and *The Crisis* were less hostile to American foreign policy than W. E. B. DuBois, Rayford Logan, and the *Pittsburgh Courier*. All criticized the Truman Administration for demanding votes for Bulgarians in Communist Eastern Europe but not for blacks in white racist South Africa; for spending billions to rescue the British Empire but not for creating opportunity for black Americans; for extracting high-profit raw materials from black Africa but not for pressing Belgium to free the Congo. To many black leaders, white racism and imperialism seemed as much a threat as Soviet Communism.

Any list of critics and criticisms of the Cold War would also include Corliss Lamont, son of the elite Lamonts of J. P. Morgan and Company and active spokesman for reconciliation with Russia through the National Council of Soviet-American Friendship. The latter body published *Soviet Russia Today*, edited by Jessica Smith, one of whose contributors was Frederick L. Schuman, Professor of Political Science at Williams College, who gave scholarly treatment to the origins of the Cold War. Another scholar, William Appleman Williams, wrote one of the first "revisionist" studies, *American-Russian Relations, 1781–1947* (1952). Scientists who were alarmed by the moral and political consequences of the atomic bomb they had helped to create became critics. Leo Szilard, James Franck, and Eugene Rabinowitch, among others, tried to prevent the use of the atomic bomb against Japan in order to avert a postwar arms race. After the dropping of the bomb, they organized the Federation of

Atomic Scientists and founded the *Bulletin of Atomic Scientists*. Albert Einstein, Harold Urey, and Linus Pauling also worked to explain the dangers of an American monopoly of the bomb and the military's control of it, as well as the need to approach Russia directly for negotiations rather than brandishing the bomb as a diplomatic weapon.

Political commentators Freda Kirchwey of *The Nation* and Vera Micheles Dean, a regular contributor to *Foreign Policy Reports*, offered tempered criticisms of and suggested alternative policies to American diplomacy in the early years of the Cold War. In 1946, Ms. Dean asked: "Is it possible that we may be expecting of Russia higher standards of international conduct than our own?"[2] Critics of varying degrees, interests, and consistency included the pacifists A. J. Muste and David Dellinger and the American Friends Service Committee; Grenville Clark, respected lawyer and United World Federalist; and Franklin A. Lindsay, executive assistant to Bernard Baruch and a member of the American delegation to the United Nations Atomic Energy Commission who, among his other contributions, wrote a perceptive memorandum in 1947 arguing against the use of loans as diplomatic pressure on Russia and Eastern Europe.

A very controversial critic was Vito Marcantonio, congressman from East Harlem. A member and leader of the small American Labor party, Marcantonio flamboyantly lashed out at American actions abroad, business and military influences in Washington, and American capitalism. Like Marcantonio, Carl Marzani, former Office of Strategic Services officer and alleged Communist spy, insisted that *We Can be Friends* (1952) with Russia. Conspicuous, but much less radical, was the dynamic director of the United Nations Relief and Rehabilitation Administration (UNRRA) in 1946, Fiorello LaGuardia. He chastised the Truman Administration for terminating UNRRA and refusing to consider LaGuardia's alternative Food Fund. The brusque, former mayor of New York City blasted Truman's "use of food relief as a weapon of foreign policy."[3] Richard Scandrett, an establishment lawyer from New York, shared LaGuardia's sentiment on relief. Scandrett served as chief of the UNRRA mission in Byelorussia and as an adviser to the Allied Reparations Commission. He scored American foreign policy for its over-simplified notions of an aggressive and totalitarian Soviet Union and particularly chided American officials for their futile attempts to win points against the Russians rather than to achieve successful reparations negotiations.

Conspicuous in his criticism of an uncompromising American stance toward Russia at the Potsdam Conference was Joseph Davies, former Ambassador to Russia. H. Stuart Hughes, a low-ranking officer in the State Department's Division of Research for Europe, has reminded us that there were some quieter, less dramatic critics of the Cold War. He has recounted that he and others in the depths of the State Department opposed unsuccessfully American self-righteousness and argued for spheres of influence because the United States was powerless to dislodge Russia from Eastern Europe. On the other hand, Elliott Roosevelt, son of President Franklin D. Roosevelt, raised a stir when he charged that Truman and his State Department advisers were shoving America away from Big Three unity, away from Roosevelt's charted peace, and were doing it "grievously" and "deliberately."[4] Henry Morgenthau, Jr., shared the view that Truman had reversed Roosevelt's foreign policy. The former Secretary of the Treasury, removed by Truman in 1945, thought Truman too rigid toward Russia. He lamented that the administration seemed intent upon rebuilding defeated Germany to the detriment of Soviet-American relations and the future peace. Finally, Norman Thomas, perennial presidential candidate for the Socialist party, considered himself to be a moderate and reasonable critic who attacked both Wall Street and Henry Wallace. Although his anti- Communism was often as virulent as Truman's, Thomas offered alternatives to the administration's military response to the Cold War.

Despite the critics' diversity and the varied intensity of their criticism, they had in common some characteristics besides their drive to find viable solutions to the Cold War deadlock. Most of them were political independents, seldom bound fast to any one party or leader and usually unpopular with major party leadership. They were loners, letting their ideas take them to the men and institutions they thought best able to extricate America from immediate crisis. Thus Pepper flirted with Eisenhower as a presidential candidate in 1948, and Taylor and Wallace left the

Democratic party for the Progressives. For the most part, the critics remained outside official Washington circles and were only occasionally listened to. Even for high officeholders such as Wallace and Taft, it was difficult to influence policy. It was Senator Arthur Vandenberg, not Taft, whom the Truman Administration invited to attend international conferences and massaged to further bipartisanship.

Most dissenters were not "isolationists" or "neo-isolationists" who supposedly were withdrawing from foreign involvements. Rather, many of them were anti-imperialists who advocated disengagement from empire, not from international relations, and less reliance on the military; most of them had favored intervention in World War II. Easily refuted too are the unsubstantiated charges that Cold War dissenters like Pepper and Wallace were simply apologists for the Soviet Union. Many of these critics, of course, possessed the fallibility of the decision-makers they criticized. Neither superhumans nor heroes, they demonstrated instances of contradiction, sometimes imagined conspiracies, often stressed one factor to the exclusion of others and hence slighted complexities, and under pressure sometimes failed to press their views vigorously. Because so few were listening to them, and because of their deep conviction, they occasionally shouted to be heard.

The Cold War critics, contrary to their detractors, were not fuzzy-minded idealists who clung to illusory hopes or who were out of touch with power politics. They recognized that postwar Soviet-American competition was unavoidable. The word "realist" is usually applied favorably to the Cold Warriors. But the word begs for clarity, because it carries at least the dual meanings of (1) tough-mindedness and (2) awareness of what diplomacy might achieve. "Tough-mindedness" defies careful measurement, and if the policies recommended by some of the critics had been followed and had worked, the realism of the second meaning would befit them. Because the critics' alternatives were rejected and Truman's policies faltered, we really do not know where realism rests. Did realism lie more in an accommodation with the Soviet Union on the basis of postwar reconstruction aid and recognition of the *fait accompli* in Eastern Europe, or did it lie in the use of economic power to coerce the

Russians and American moves in Eastern Europe to challenge the Soviets in their own backyard? The American Cold War debate cannot be explained simply as a clash between realists and nonrealists.

The long-accepted idea that the Cold War was inevitable and that the United States could have done nothing to change it has perhaps obscured the history of its critics more than any other factor. It has also allowed adherents to the interpretation of inevitability to give scant attention to the critics' alternative proposals and to assume that there were very few choices before official Washington at that time. In this view, it follows that the Truman Administration made decisions which were largely forced upon it from abroad, especially by an obstreperous Russia, and that the dissenters' alternatives were irrelevant. It suggests that American policymakers were not free to change events—did not have the chance or power to make different decisions. The complexity of Cold War history and the significance of American contributions to the crises of the Cold War are thus overlooked for a heavy emphasis on Soviet intransigence, Stalin's brutalities, Communist ideology, and an obstructionist international Communist conspiracy. The historian Arthur M. Schlesinger, Jr., for example, even after surveying some of the errors and stubbornness of American foreign policy, easily shifted in a 1967 article in *Foreign Affairs* to argue that what really counted, what really brought on the Cold War, were messianic Leninist ideology and totalitarian, Stalinist paranoia, neither of which the United States could do anything about and before which it was largely helpless. But the Cold War was surely not so cut-and-dried, foreordained, or inexorable, and, as the critics courageously demonstrated, American decisions helped spawn international crises.

The critics' alternatives ranged from specific proposals relating to single issues, such as Germany and the Greek civil war, to more general calls for world federalism. There were recommendations on the atomic bomb and its control, international organizations, foreign aid and trade, the Truman Doctrine, Germany and its postwar occupation, the Marshall Plan, NATO, colonialism, Eastern Europe, China, internal security programs, and the Korean War. Some of the suggestions were simply the negatives or the opposites of declared policies: Do not abandon UNRRA. Do not rearm Germany. Avoid the simplistic and alarmist language of the Truman Doctrine. Do not attempt to use the atomic bomb for diplomatic bargaining. Do not support European imperialism in Africa and Asia. In their alternatives and criticisms, the critics exposed an international double standard applied by Washington and American manipulation and sidestepping of international institutions. They pinpointed uncompromising American positions which obstructed negotiations. They attempted to allay exaggerated and sometimes hysterical American fears of Communism and stressed the dangers of the Cold War to domestic well-being.

Most of those critics who ardently bucked the bipartisan consensus found themselves recipients of calculated distortion and charges ranging from idiocy to conspiracy. The 1947 memorandum Clark Clifford prepared for the President, explaining how Truman could defeat the upstart Wallace, bears repeating: "Every effort must be made *now* jointly and at one and the same time—although of course, by different groups—to dissuade him and also to identify him and isolate him in the public mind with the Communists."[5]

Attempts to tag critics with Communist labels were popular long before Joseph McCarthy brought ostentatious drama to the practice. The Truman Administration itself, employing a language of immediate crisis and fear to move its programs through Congress and to deflect public criticism, helped prepare the way for McCarthy's more explosive diatribes. It was Attorney General J. Howard McGrath in 1949 who said that "there are today many Communists in America. They are everywhere—in factories, offices, butcher stores, on street corners, in private businesses. And each carries in himself the germ of death for society."[6]

Pepper, Taylor, and Wallace were vilified as Soviet apologists, the Council of African Affairs was placed on the Attorney General's subversive list, and W. E. B. DuBois was indicted in 1951 for refusing to register as an agent of a foreign government (Marcantonio successfully defended DuBois). Conservative critics such as Taft, on the other hand, were belittled as "isolationists." Defenders of official policy refurbished the words "appeasement" and "Munich" to disparage alternative proposals. Wallace understood the effectiveness of such attitudes and treatment in 1948, when he gathered but 1,157,000 votes or only about 2 percent of the total cast. The mild critic Richard Scandrett summarized the plight of dissenters in 1948: "Even to suggest in a whisper here nowadays that every Russian is not a cannibal is to invite incarceration for subversive activities." Scandrett lamented that "that is the American way. Things are either pure black or pure white. There are no greys."[7]

The favorable popular response to the Marshall Plan as an example of American generosity and selflessness, and the widespread and horrified American reaction to the 1948 Communist coup in Czechoslavakia and the Berlin Blockade, further undermined dissent in the escalating Cold War. The crushing blow for many critics was the Korean War. This war, which Truman read as a clear-cut case of Soviet aggression, fit the "we-told-you-so" category. Many critics met defeat at the polls in 1950 or succumbed finally to the Cold War consensus. Pepper lost his Senate seat after a vicious campaign by George Smathers. Taylor and Marcantonio met a similar fate. Wallace and many other critics endorsed American intervention in Korea. Under considerable and unrelenting pressure, some of the critics crumbled under the weight of the hostility against them. The power of the consensus that Truman had shaped was extensive, and it eventually forced many dissenters to fall in line.

Most of the critics made little immediate impact upon American foreign policy. Some were harassed and dismissed as misguided aberrations; others received polite if cool attention. For the most part their alternatives were not given careful hearing. Yet their questioning, vigilance, and suggestions were checks of varying degrees upon the Truman Administration in the first years of the Cold War. Senator Taft's critique of NATO, according to Senator Vandenberg, "has given me a first class headache. . . . "—that is, a longer debate on the important issue than either Vandenberg or Truman wanted.[8]

One critic who gave repeated headaches to Truman officials was the junior senator from Florida, Claude D. Pepper. His story provides a case study of the perils of Cold War dissent and the strength of the Cold War consensus about the Soviet threat. In his successful bid in 1950 to unseat Pepper, George Smathers's campaign invidiously published *The Red Record of Senator Claude Pepper*. Some of Pepper's statements isolated from context were arranged to suggest that he was a Communist. His loudest opponents had often accused him of flirting with Communism, befriending Josef Stalin, practicing appeasement, and serving as a leading apologist for Soviet foreign relations. Clark Clifford informed Truman in 1947 that Pepper was "a devout if cynical follower of the Party line," and the *New York Times* fed the popular image of "Red" Pepper by reminding its readers that *Pravda* and Russian officials applauded some of Pepper's speeches.[9] Smathers opened his campaign on January 12, 1950: "The leader of the radicals and extremists is now on trial in Florida. Arrayed against him will be loyal Americans. . . . Florida will not allow herself to become entangled in the spiraling web of the Red network. The people of our state will no longer tolerate advocates of treason."[10]

Claude Pepper was only thirty-six years old when Floridians sent him to the Senate in 1936. Behind him were a rural Alabama childhood, jobs as a hat cleaner and steelworker in Birmingham, an active life of oratory and politics at the University of Alabama, years of insecurity at Harvard Law School, and a one-year stint as instructor of law at the University of Arkansas where J. William Fulbright sat as one of his students. In 1925 the young lawyer was attracted to Florida by a short-lived land boom. Pepper practiced law for a few years, in 1928 was elected to the Florida House of Representatives, and eight years later won his United States Senate seat as a strong advocate of New Deal programs. As the Second World War approached, Pepper incurred the wrath of isolationists for his advocacy of aid to Great Britain.

In the postwar period, Pepper argued that the Truman Administration had abandoned Roosevelt's foreign policy of international cooperation for a "get-tough" policy surfeited with a double standard, anti-Communist hysteria, unquestioning consensus and bipartisanship, and American unilateralism. Pepper's political maneuvers (he announced his own candidacy for the presidency in 1948, then reluctantly withdrew to support

Truman) and his criticisms of Truman policies made him anathema to official Washington. The leadership of the Department of State disliked him because he had questioned the appointment in 1944 of a number of assistant secretaries whom he considered to be representatives of wealthy interests. The President remembered that Pepper had favored Wallace over him for Vice President in 1944, and that Pepper continuously upbraided him for waylaying the New Deal. As Truman put it later, Pepper was an "agitator."[11] Secretary of State James F. Byrnes told a Cabinet meeting that "his negotiations . . . were greatly embarrassed by the public utterances" of Pepper and Wallace.[12] Pepper's party and Senate colleagues punished him for upsetting the foreign policy consensus by keeping him off the Senate Foreign Relations Committee in 1947–48. After his vote against the Truman Doctrine, a publisher canceled his book contract.

Many conservative voters in Florida were alarmed by Pepper's failure to participate in Southern filibusters, his opposition to the poll tax, his support for an anti-lynching law, and his advocacy of a national health insurance plan tied to social security. He called for an Equal Rights Amendment and sponsored minimum-wage legislation. When Truman startled Congress and the nation during the railroad strike of 1946 by asking for power to draft into the armed services striking workers of any industry taken over by the government, Pepper joined Senator Taft to argue against the dangerous bill. Pepper also voted against the Taft-Hartley Act. Pepper remained firm in his belief that "if a man really means his liberalism, he will make enemies."[13]

Pepper invited many enemies, but at times he stood with the majority in his party and in the Senate. He voted for Truman's domestic programs. In foreign policy, embracing the idea that international economic collaboration was essential to peace, Pepper supported the Reciprocal Trade Agreements Act, the Bretton Woods Agreements (World Bank and International Monetary Fund), and the $3.75 billion loan to Great Britain (1946). He was an enthusiastic champion of the United Nations Organization. And he shared the pervasive American concept of peace and prosperity—that is, that a healthy world economy and international trade were necessary foundations to pacific international relations, and that American foreign trade was vital to both American and international prosperity.

It was Pepper's vigorous criticism of American foreign policy that distinguished him and consumed most of his time. He was convinced that a peaceful accommodation between the Soviet Union and the United States was possible in the critical 1945–47 period. To explain the heritage of hostility, Pepper often outlined the history of Soviet-American relations and the long-standing tension fed in part by Soviet fears for its security and social system. He recalled American and Allied intervention against Bolshevism in the Russian Revolution and the post-World War I economic blockade, both of which left diplomatic scars. Soviet Russia was excluded from the League of Nations, belatedly recognized in 1933 by the United States, and ostracized from the Munich Conference of 1938. Pepper concluded that America in the postwar period was inflaming old Soviet fears by denying the Soviets a sizeable loan like that granted to the British, by constructing foreign military bases, by monopolistic control, stockpiling, and new testing of the atomic bomb, by domestic anti-Communist fervor, and by support for anti-Russian governments such as the London Polish exile group and Franco's Spain. And the Russians could not ignore, Pepper argued, America's bypassing of the United Nations, the rebuilding of Russia's archenemy Germany, and the use of American economic power through loans, trade, and relief to mold an anti-Soviet bloc. Pepper recognized that Stalin's Russia was no international saint, but suggested that some of Russia's uncooperativeness derived from its response to American actions and the American application of a double standard.

Pepper spoke frankly on the origins of the Cold War: "I think one of the things that contributed to the breach that is now giving us the cold war," he told a Chicago audience in 1948, "was the fact that they got the impression immediately after World War II that they were not going to have any aid from us, and they would have to make it on their own."[14] Indeed, the American decision to deny postwar reconstruction aid to Russia led to an early diplomatic roadblock. Senator Pepper was closer to this subject than most of his colleagues, because he discussed postwar aid with Stalin in Moscow during a European tour in the fall of 1945. Stalin catalogued Russia's massive war damage and appealed for heavy equipment and food from the United States. In return for American products, Russia would offer valuable manganese ore, chromium ore, timber, and gold. The Russian leader expressed "chagrin" that America had not opened negotiations on the Russian request for a loan, and stated that it would be "suicide" for the Soviets to use the funds for military purposes in the face of such economic dislocation.[15]

The history of the abortive American loan to Russia suggests that Pepper's advocacy of postwar aid to Russia was a viable alternative to American diplomacy in the 1943–45 period. Secretary of State Edward R. Stettinius, Jr., later wrote that the loan issue was "one of the great 'if' questions of history."[16] Ambassador W. Averell Harriman himself cabled Washington in early 1945 that the Soviet Union placed "high importance on a large postwar credit as a basis for the development of 'Soviet-American relations.' I sensed from his [V.M. Molotov's] statement an implication that the development of our friendly relations depends upon a generous credit."[17] But the Truman Administration soon developed a policy of toughness which impeded any movement toward negotiations. The United States insisted that the Soviet Union create an "open door" in Eastern Europe and disclose its trade treaties with the nations of that area. The Russians replied that such treaties were not negotiable with the United States. Nor would the Russians discuss membership in the World Bank, an American-dominated institution from which they could derive little benefit. The Truman Administration attempted to use the loan as a diplomatic weapon before negotiations, rather than as a diplomatic tool during discussions.

The abortive loan issue raised some provocative questions. Would the Soviet Union have sought such heavy reparations from former Axis nations in Eastern Europe and would the issues arising from that region have been so tension-ridden had a loan been granted? Harriman concluded in November 1945 that American economic policy toward the USSR had "so far added to our misunderstanding and increased the Soviet's recent tendency to take unilateral action." He added: "This has no doubt caused them to tighten their belts as regards improvement of the living conditions of their people and may have contributed to their avaricious policies in the countries occupied or liberated by the Red Army."[18] Would the Soviet Union have been so demanding vis-à-vis Germany had a loan been granted? Would the Soviets have eased their reparations demands and agreed early to unite the German zones if the United States had acted with speed on the loan? An assistant to Donald Nelson concluded that "it seems altogether probable that these two matters, an American credit and German reparations, were closely linked in Soviet political thinking, for our attitude toward both questions profoundly affected the rate of Russia's postwar recovery."[19] Indeed, Edwin Pauley, American Reparations Ambassador, surmised that "Russia's intransigent position on unification and reparations is due to a desire to obtain the maximum amount of industrial and consumer goods from Germany, to meet internal political prestige needs and to rebuild the Soviet industrial machine."[20] Pepper was a lonely figure in the Senate on this question, reminding that body that the denial of a loan in 1946 increased Soviet suspicion and fear.

If the loan denial exacerbated Soviet-American conflict, so did other American decisions from which Pepper dissented and to which he presented alternatives. The postwar trend of building an American military establishment abroad, the American monopoly of the atomic bomb, and the growing alliance system alarmed Pepper. The "get-tough" approach to diplomacy, Pepper warned, was a prostitution of Roosevelt's cooperation with Russia and his policy of holding out "the friendly hand of the good neighbor" abroad.[21] Toughness in diplomacy would only encourage other powers to acquire "bigger sticks, and we begin that never-ending vicious cycle of measure and counter-measure."[22] The major powers, he wrote, "might as well recognize that they cannot indefinitely maintain the mastery of the strategic defense areas of the world without Russia striving for a comparable position."[23] He called for the removal of Soviet troops from Hungary and Rumania, but also asked the United States to rescind its decision to construct an air base in Iceland and refrain from fleet demonstrations in the Mediterranean. Pepper also urged Truman to invite Britain to withdraw its troops from Greece, Iraq, Iran, and Jordan, and to give up its colonies. "What I decry is the international hypocrisy, sham, and pretense. If the British people want the Russians to get their troops out of Iran, let the British get their troops out of Iraq."[24] Pepper feared that America would become a guarantor of British imperialism.

Senator Pepper, like many Americans, sought an end to the nuclear arms race among the great powers. As long as America continued to stockpile, test, and maintain a monopoly of atomic weapons, Pepper concluded, it was to be expected that Russia also would strive to develop the destructive instruments. He chastised those who "choose to think of atomic power not as a boon to mankind, but as a weapon America can hold over the rest of the world to force it to remake itself in our image and to our profit."[25] The Truman Administration and Pepper clashed sharply over the means to achieve international control of the bomb and atomic energy. Pepper stated simply that the United States should "destroy every bomb we have, and smash every facility we possess which is capable of producing only destructive forms of atomic energy." With this action, Americans could enter international discussion "with the cleanest of hands. . . ."[26] Knowledge of the bomb's construction should also be turned over to a committee of the United Nations. Pepper did not offer a careful policy alternative, and his appeal for destruction of the bomb was naive; it was visionary to expect any nation to divest itself painlessly of a superior military weapon.

But so, too, was the Baruch Plan of 1946 visionary. Proposed by the Truman Administration, the plan would have established an International Atomic Development Authority empowered to control the use and sale of all fissionable materials. An inspection system would be set up, and America would relinquish its control of the bomb only after the entire system was functioning. The Russians rejected this scheme, finding it too protective of the American monopoly and an infringement upon Russian sovereignty. Henry Wallace regretted that "we are telling Russians that if they are 'good boys' we may eventually turn over our knowledge of atomic energy to them and to other nations."[27] The United States asked the Soviet Union to open its territory, its military secrets, and its resources to an international investigating team, to avoid producing or developing the bomb, and to accept an international commission where the veto was forbidden and a pro-American majority would sit—all before America gave up its exclusive control of the weapon. Senator Pepper offered an unconvincing solution, but his diagnosis was accurate: America could not confine Russia behind a curtain of atomic weapons and retain mastery of such armaments without arousing Soviet resentment and encouraging an international arms race.

When, in March 1947 President Truman requested direct American military and economic aid for Greece and Turkey, Pepper joined Senator Taylor to offer a dissenting proposal in the form of a joint resolution. They agreed with Truman that war-exhausted Greece needed relief and economic recovery; indeed, Pepper had urged aid to Greece as early as March 1945. But he and Taylor argued against any military aid to Turkey. Turkey was ruled by a dictatorship, hardly fitting Truman's claim that America was building democracies in the Near East, and had more often than not supported Germany in World War II. Russia was not militarily threatening Turkey; rather, it was demanding joint control over the strategic Dardanelles—a traditional Russian desire more than a Communist one. Stalin stated at the Yalta Conference that he could not "accept a situation in which Turkey had a hand on Russia's throat."[28] The United States refused to discuss joint Soviet-Turkish control after World War II and encouraged the Turks to be uncompromising. Pepper demanded that a double standard not be applied. The Dardanelles were to Russia's defense what the Gulf of Mexico and the Panama Canal were to that of the United States. America, Pepper emphasized in the Senate, should not presume to determine the arrangement of control in the Dardanelles, unless it was willing to invite Soviet consultation on the Caribbean and Panama. Pepper thus recognized the realities of international geopolitics and existing spheres of influence.

Besides denying military aid to Turkey, the Pepper-Taylor alternative program suggested that the United Nations administer an American contribution of $250 million for non-military uses in Greece. Personnel of the United Nations Relief and Rehabilitation Administration were still in Greece and they might be employed, and a United Nations investigating team was already there. The Food and Agriculture Organization (FAO) of the United Nations issued a 1946 report advising a relief program of $100 million. The Truman Administration nonetheless ignored the FAO, a body not subject to veto authority by any of the major powers. Acheson admitted to Pepper that the United States at no time had consulted the United Nations or any of its members about the possibility of a United Nations program. Indeed, Dean Rusk, Director of the Office of Special Political Affairs which handled United Nations relations in the State Department, was not even consulted during the development of the Truman Doctrine. Senator Pepper with Walter Lippmann feared that this unilateral American action would weaken severely the United Nations.

Pepper noted in his critique of the Truman Doctrine, and most historians now agree, that the Greek monarchical regime was reactionary and hostile to social change, that the civil war began during the Second World War, that the British had tried desperately to shore up the faltering regime against the rebels, and that many revolutionaries were not Communists but rather Greek nationalists. Furthermore, the civil war was not engineered or carried out by the Soviet Union. "I do not support the thesis that everywhere there is a Communist he was sent there by the Russians, and is directed and probably paid by the Russians," Pepper asserted.[29] In the United States, however, the popular and mistaken assumption prevailed that Russia was manipulating the Greek revolution. What American intervention meant, concluded Pepper, was a "stepping into empty British footprints in the imperial quicksands of the world." American military personnel in the Near East would be the "emissaries of a new imperialism," and the United States could not then "expect to carry conviction in our condemnation of the Russians for doing some of the same things."[30] Pepper cast a nay vote, but the Senate approved aid to Greece and Turkey by a 67 to 23 count.

During 1948 Pepper gradually shifted to support the Truman Administration's foreign policy. He endorsed the Marshall Plan. He was shaken by the Communist coup of February in Czechoslovakia. To him the event meant more than the elimination of an independent government; it meant also that his warnings that a unilateral "get-tough" American diplomatic stance would heighten world tension had been proven correct. The collapse of Jiang in China further shocked Pepper, and the outbreak of the Korean War prompted him to issue a scathing attack upon the "Russian onslaught," and to applaud the action of the United States-United Nations effort to hurl back aggression. "Hell is not hot enough for those Red criminals."[31] Disturbed by events in Czechoslovakia, China, and Korea, aware of growing opposition to his re-election in Florida, stung by Smathers's redbaiting, and burdened by a sense of defeatism and helplessness, Pepper began to vote for such measures as NATO, although he hoped the Atlantic pact would be limited to a few years.

The Cold War consensus had ultimately conquered and molded Senator Claude Pepper. He became despondent in 1949 when he recalled that, in the first years of the Cold War, "I preferred another course. But my views did not prevail."[32] The aggression and bellicosity that was absent from Soviet foreign policy in the period 1945–47 was there in 1948–50, Pepper had come to conclude. Almost alone in the Senate, Pepper had appealed for a consistent standard, mutual cooperation, economic reconstruction, utilization of the United Nations, and international conferences. He believed that both Russia and the United States were responsible for the coming of the Cold War. He urged Americans to cast off their exaggerations of the Soviet threat. After persistent setbacks and vituperative attacks upon his views and integrity, Pepper's spirit broke. Overcome in 1950 by the Cold War, anti-Communist mania, social and political conservatism, and white racism, Pepper in his political defeat became a victim of the American penchant to exaggerate the Communist threat.[33]

# 7

# From Architect of Containment to Critic: George F. Kennan and the Soviet Threat

In February 1946 Washington eagerly sought an interpretive report on Stalin's election speech and Moscow's reasons for not joining the World Bank—in short, what made Soviet foreign policy tick? For years George F. Kennan had been preparing himself for just such a request. "The more I thought about this message, the more it seemed to be obvious that this was 'it.'" He was tired of plucking people's sleeves to draw their attention to the Soviet menace. "Here was a case where nothing but the whole truth would do. They had asked for it. Now, by God, they would have it." On February 22, Kennan completed his "long telegram," an eight-thousand word "elaborate pedagogical effort," which ultimately elevated him to the heralded status of *the* expert on the Soviet Union and *the* intellectual in the Foreign Service.[1]

The telegram began with a discussion of the "basic features" of the Soviet outlook on international relations. Placing heavy emphasis on ideology as the wellspring of Soviet behavior, Kennan pointed out that the Soviet mind was geared to capitalist-socialist conflict. Hostility to the capitalist outside world was not a Russian attitude, but a Communist one, a party line. Yet at the bottom of the Kremlin's "neurotic view of world affairs is traditional and instinctive Russian sense of insecurity."

Marxist-Leninist dogma, argued Kennan, was the new vehicle for perpetuating this Russian tradition. The Soviets after the war, he predicted, would use Communist parties as an "underground operating directorate of world communism." Russia would seldom compromise; in fact, its foreign policy would be "negative and destructive in character, designed to tear down sources of strength beyond the reach of Soviet control." In summary, "we have here a political force committed fanatically to the belief that with US there can be no permanent *modus vivendi*, that it is desirable and necessary that the internal harmony of our society be destroyed, the international authority of our state be broken, if Soviet power is to be secure."[2] Kennan closed his message with less alarmist language. He believed Russia would not risk a major war because it was too weak. America, furthermore, could counter Soviet expansion by constructive propaganda efforts at home and abroad and by maintaining a healthy and cohesive society in the United States as a model. Kennan seemed to emphasize that Russia was more a "political" than a "military" threat, but his statements could be variously interpreted.

Officials in Washington read the long telegram with relish and excitement, for Kennan seemed to have captured their own moods and fears. Here was a document that was authoritative, convincing, and simple, demonstrating just how intransigent the Soviets were. Kennan's timely message helped persuade the administration to speak out more forcefully against Soviet foreign policy, to "get tough," in popular jargon. It had actually been following a "tough" policy before, but now it was more willing to follow it publicly. Truman Administration figures chose to emphasize the parts of Kennan's message that described an uncompromising, warlike, aggressive, neurotic, and subversive Russia that had to be stopped. Whether Kennan intended to or not, he fed exaggerated fears that an international Communist conspiracy, directed by Moscow, sought to subvert the world. He later confessed that the telegram read "exactly like one of those primers put out by alarmed congressional committees or by the Daughters of the American Revolution, designed to arouse the citizenry to the dangers of the Communist conspiracy."[3] As the Cold War progressed into the 1950s,

Kennan refined his thinking, regretted his own hyperbole, and became a critic of an American foreign policy whose assumptions and course he himself had once helped shape.

Scholars consider him a first-rate historian; Foreign Service officers rank him as one of their best; Presidents and Secretaries of State have tapped his expertise on the Soviet Union; college students have been weaned on his publications and prescriptions for a "realistic" foreign policy; and literally hundreds of thousands of Americans and foreigners have become familiar with his contemporary analyses of international relations. Soviet leaders went so far as to declare him *persona non grata* in 1952. Critics of the war in Indochina in the 1960s welcomed him as an ally, but dissenters from American foreign policy since World War II identify him as the masterful Cold Warrior who helped launch the global containment doctrine. In the 1970s and 1980s he distinguished himself as an analyst of the nuclear arms race and its threat of doomsday. George Frost Kennan has pursued a demanding and varied career as diplomat and intellectual, often at the center of policymaking and public controversy; yet he has claimed for himself little influence over events, ideas, or individuals. Indulging his habit of self-anguish, Kennan has always underestimated his influence upon a generation of Americans.

Throughout his career Kennan lived with two competing aspirations. One was the life of the bureaucrat, the Foreign Service officer, legalistic and precise, dedicated to a professional creed and style, enamored with tradition, distant from the winds of domestic politics, disciplined, moderating the opinions of the overzealous, and always eager to influence decision-makers. Kennan liked being an "insider," for that role provided a needed sense of belonging, prestige, and access to power. Yet he never seemed comfortable in the foreign affairs bureaucracy; he found it too confining and encumbered, too tied to routine and menial tasks, too demanding of subservience to assigned missions. More than once he complained that his superiors, usually depicted as less able than himself, ignored his talents. He became chagrined that Foreign Service Officers found themselves "tidying up the messes other people have made, attempting to keep small disasters from turning into big ones," instead of contributing fresh policy perspectives.[4] He resented the intrusions of uninformed congressmen and a whimsical public and bemoaned the decline in stature and power of the Department of State. It seems fair to conclude that Kennan has felt frustrated because for most of his career he sought but was denied an influential role, to use a recent example, like that given to Henry A. Kissinger by President Richard M. Nixon.

Kennan's other aspiration propelled him toward the life of the intellectual, the scholar playful with ideas, speculative, independent of mind, skeptical and probing, the dedicated teacher, the moralist preaching his convictions, the oracle of reasoned vision, and the artist. This was a life of individualism and loneliness, opposed to the ordinary and routine, always suspicious of pat answers. Whereas the diplomat-bureaucrat attempts to be precise and methodical, the intellectual muses and experiments. Kennan's "irrepressible intellectual brashness," as he described it, found sustenance in the role of the "outsider."[5] The life of the intellectual also had drawbacks for Kennan, not the least of which was isolation from the corridors of power and the awful possibility that his ideas would be misinterpreted. After the 1920s, when he entered the Foreign Service, Kennan essentially pursued two careers simultaneously, not finding complete fulfillment in either. Therein lies his self-image as a failure and the differing interpretations others have given to his conception of Cold War diplomacy, as well as his ultimate alienation from the government he sought so faithfully to serve.

Throughout his life, Kennan nurtured remarkably consistent attitudes and values; as he himself observed, they were quite different from those of the Soviets, whose ideology and government he studied so closely. He has described himself as a "conservative person, a natural-born antiquarian, a firm believer in the need for continuity across the generations in form and ceremony. . . ."[6] Indeed, he looked to the past for models after which to pattern ideas and behavior. Because he witnessed the deterioration of his revered values, he never seemed at home in the twentieth century (or in his own country). Words like honor, responsibility, efficiency, rigor, discipline, steady habits, perseverance, self-restraint, individualism, culture, tradition, decency, courage, cleanliness, order, and humility mark

Kennan's reflection on himself and on others—as if he had torn a page from Benjamin Franklin's autobiography. These qualities added up to something he vaguely called "Western civilization," rooted in morality, pride in work, competition, individualism, freedom of thought, and social conservatism. In contrast, Marxism and the Soviet Union prostituted these values through collectivism, fanaticism, denial of individual rights, and a break with tradition. In the early 1930s, Kennan wrote that he "rejected the Communists because of their innate cowardice and intellectual insolence. They had abandoned the ship of Western European civilization like a swarm of rats when they considered it to be sinking, instead of staying on and trying to keep it afloat. . . ." The Communists "had called all their forefathers and most of their contemporaries hopeless fools," and that behavior was a "sacrilege." "I felt," Kennan explained, "that it [Communism] must some day be punished as all ignorant presumption and egotism must be punished."[7]

He visited his arrogance and contempt upon Americans as well. He lacked patience with people less bright than himself and held little respect for a political system that permitted the untutored to vote. He chided Americans for their lack of historical perspective, low level of knowledge about the country's problems, and emotionalism. Frequently emotional himself, he could not tolerate it in others, especially when in collective form—in the "masses." He seldom concealed his disdain for those who neither aspired to nor recognized his own quest for the attributes of the ideal diplomat: charming, brilliant, cultured, educated, self-confident, and respected. For much of his life he seemed out of touch with the country he served abroad. He observed and understood foreign countries much better than his own, and, when in the early 1950s he returned to private life, what he saw in America repelled him.

Style and method were particularly significant to Kennan. It was crucial *how* one did what one had to do. "Let no one underestimate," he remarked, "the importance in this life of the manner in which a thing is done. It is surprising how few acts there are, in individual life, which are not acceptable if they are carried out with sufficient grace and self-assurance, and above all with dignity and good manners. . . ."[8] Kennan was easily put off by the style of others, from the often rude, blunt Soviets in the 1930s and 1940s to the long-haired iconoclasts of college campuses in the 1960s. Kennan revealed a bureaucrat's obsession with style and routine and an aristocrat's concern with propriety.

These traits were evident long before Kennan emerged as a major architect of Cold War diplomacy. He was born in 1904 to an Anglo-Saxon family in Milwaukee, Wisconsin. His father was a hard-working lawyer and his mother a self-conscious "aristocrat." His family had no quarrel with a society that had rewarded it with a more than adequate income, and Kennan has reflected that his early experience refuted the Marxist image of the blood-sucking capitalist. "I could identify neither with the exploiter nor with the exploited."[9] At age twelve he headed for St. John's Military Academy near Milwaukee. Its strict discipline and Spartan environment left little time for frivolous pleasures, and Kennan learned to keep his room neat and orderly and to merge his personal preferences with those of the group. When he graduated in 1921, as if to illustrate that other aspiration, he was designated class poet.

A St. John's administrator advised the somewhat shy Kennan to go to Princeton University, an inclination already stimulated by his reading of F. Scott Fitzgerald's *This Side of Paradise* (1920), the novel set in Princeton. He was, he later wrote, "an oddball on campus, not eccentric, not ridiculed or disliked, just imperfectly visible to the naked eye." He became a loner, knowing few people, distant from his instructors, always last in line, with no desire to join one of the university's prestigious clubs. Later he recalled that he sought a kind of "martyrdom" in those years, and in 1925 left college "as obscurely as I had entered it."[10]

Not knowing what else to do, Kennan decided to apply to the newly formed Foreign Service. He had liked his limited study of international affairs at Princeton. And he had always been impressed with the record of his grandfather's cousin, also bearing the name George Kennan, who had earned a reputation as an expert on Russia. Then, too, Ivy League graduates predominantly staffed the diplomatic corps at that time, so Kennan was following a path well traveled by Princetonians. After a rigorous oral examination conducted by Under Secretary of State Joseph Grew, Kennan was accepted in 1926 into the Foreign Service School. He then spent several months in the consul general's office in Geneva, shedding his "introverted" self for that of a man of responsibility in a noble profession. He next served as a vice-consul in Hamburg, where he became absorbed in the intellectual life of that Social Democratic city, attending lectures and musing beside the famed harbor. But, apparently stifled by his job, he decided in 1928 to return to the United States to undertake graduate study.

His superiors in the Foreign Service induced him to change his mind by offering him three years of graduate work in Europe, if he would study the Arabic, Chinese, Japanese, or Russian languages. Kennan chose Russian, in part because the elder George Kennan's "tradition" beckoned him, in part because he sensed that the United States would eventually recognize the Soviet Union and need experts. Area specialization was something new for the Foreign Service, and Kennan became one of a handful (including Charles Bohlen and Loy Henderson) trained in Russian affairs in the 1920s. Already fluent in German and French, he departed for Estonia to study Russian. He learned fast and was assigned to Riga, Latvia, America's contact with Russia, where travelers to and from Russia were interviewed and Soviet publications collected. From Riga he was sent to the University of Berlin to study Russian history, literature, and language. During the two years there he became annoyed with American "liberals" tolerant of the Soviet experiment, because, as he wrote in 1931, "the present system of Soviet Russia is unalterably opposed to our traditional system, that there can be no possible middle ground or compromise between the two. . . ."[11] This pessimistic view was probably encouraged by his contact with anti-Soviet Russian émigres.

Assigned again to Riga in 1931, now third secretary of the legation, Kennan eagerly threw himself into his work, and from his desk flowed economic reports, analyses of Soviet propaganda, and even an article, "Anton Chekhov and the Bolsheviks." He was less enthusiastic than many Americans about the prospects for Soviet-American commerce, because the Soviets would need American credit, and he advised that diplomatic recognition of the Soviet Union would not be a panacea for facilitating trade. Kennan was not pleased with the imprecision and irresolution of the Roosevelt-Litvinov Agreements of 1933. Later he chided the President for thinking that recognition would somehow curb Germany or Japan. As Kennan recalled, "never—neither then nor at any later date—did I consider the Soviet Union a fit ally or associate, actual or potential, for this country."[12]

On leave in Washington during the Litvinov-Roosevelt talks, Kennan was introduced to William C. Bullitt, soon to be the first American Ambassador to the Soviet Union. They established an immediate rapport, and Bullitt, impressed by Kennan's knowledge of the Russian economy and his fluency in the language, added Kennan to his staff. Kennan probably concealed his pessimism about Soviet-American relations from the more optimistic ambassador. The initially cordial Soviets were delighted with the young diplomat because they identified him with the anti-czarist elder Kennan and because his mastery of Russian flattered them. The sluggishness of the Soviet bureaucracy exasperated Kennan, but the embassy staff, which included the journalist-diplomat Charles W. Thayer and the career officers Henderson and Bohlen, exhilarated him. In 1935 Russia entered the terrible period of the purges, and Kennan became an acute and outraged observer, personally attending many of the trials. It became a "liberal education in the horrors of Stalinism." Stalin had eliminated "liberal" forces in favor of a "new gang of young and ruthless party careerists," who had abandoned "modern western civilization" for conspiratorial thinking, secretiveness, cruelty, unscrupulousness, and opportunism.[13]

Kennan had no liking for the "proletariat" or for Marxism. His elitism and conservatism were apparent when he took a short trip to a Black Sea resort area in 1936. He was disgusted with the "simple people's" lack of grace; they made "pig sties" of the hotels and lacked appreciation for the scenery.[14] For Russian Marxism there was nothing but intellectual distaste. He rejected its "stony-hearted fanaticism" and its contribution to the liquidation of groups of people like the bourgeoisie because of a rigid application of class distinctions. The Marxist "ideology" was "pseudo-science, replete with artificial heroes and villains. . . ."[15] The new restrictions on diplomats during the purges exacerbated Kennan's intense disaffection. An aura of suspicion gripped Moscow, and Soviet officials tried to isolate foreigners from the Russian people, with whom Kennan had built up some rewarding relationships.

During the 1930s Soviet-American relations deteriorated over the failure to reach agreement on the payment of First World War debts, the purges, and the meeting of the Comintern Congress of 1935. As Bullitt soured on the Soviet Union and moved toward the "hard line" so much appreciated by Kennan, the two grew close in their thinking. Bullitt resigned in 1936, and Kennan considered his replacement, Joseph E. Davies, a political hack, ill-fitted for the post, lacking in seriousness, looking too much after his public image. It was an exaggerated assessment, but Kennan chastised Franklin D. Roosevelt more than Davies, for the President was a politician and inadequately hostile to the Soviet Union. Kennan contemplated resigning from the Foreign Service, but chose instead to make the best of a bad situation and departed for Washington in 1937.

Kennan accepted the Russian desk post in the Department of State and for about a year talked with businessmen, protested Soviet restrictions on the Moscow embassy, and watched for violations of the Litvinov-Roosevelt Agreements. It was not a happy time for him—he felt ignored and unappreciated—but his ideas about Soviet-American relations jelled further. The Russians, he concluded in 1938, had been influenced by "Asiatic hordes" during their long history and were thus conditioned to think of foreigners as enemies. The Asiatic's "acute and abnormal sense of 'face' and dignity" compelled Russian leaders to claim infallibility. Furthermore, the size of the country contributed not to a sense of moderation, but rather to "extremism." The Russian character took on the traits of a "typical oriental despotism" and at the same time an "inferiority complex," because the Russians were aware of their own defects. Kennan called for "patience" in American foreign policy: "We must neither expect too much nor despair of getting anything at all."[16] His conclusions were quite deterministic and subjective, but he spoke with amazing assurance. Hostility toward and deep suspicion of the Soviets, as well as the counsel of patience, became the hallmarks of Kennan's thought.

In mid-1938 Kennan was assigned to the American legation in Prague. When German troops entered that somber city in March 1939, the post was shut down, and he moved to the embassy in Berlin. He seemed to accept the fall of Czechoslovakia as predestined, because he believed the breakup of the Austro-Hungarian Empire after World War I had been a mistake, fragmenting power in the Danube Basin and inviting international squabbles. In Berlin he watched Europe go to war. When Germany invaded Russia in June 1941, Kennan approved Lend-Lease aid in America's self-interest, but warned against professions of friendship with or moral support for the Soviet Union, which he considered hardly a fit ally. From December 1941 to May 1942, the Germans interned Kennan and about 130 other Americans. Kennan assumed leadership and maintained discipline, but became bitter toward Washington, believing his government had abandoned the group. After the State Department finally gained his release, he became deeply annoyed when half of his group was ordered to wait for another ship so that Jewish refugees could be transported to freedom first.

Back in the United States, he rested and thought. He soon identified himself as a moderate between extreme anti-Soviet and pro-Soviet factions in the United States. In apparent anticipation of considerable Soviet influence in Eastern Europe after the war and in apparent contradiction to his stated views against self-determination for small nations in the area, he opted for the latter in the hope of curbing Soviet expansion. Before the war he demonstrated no such strong opposition to German expansion in that region and argued that small states in fact invited instability. In September 1942, Kennan was assigned to Lisbon as counselor. He did not go willingly. He complained to the State Department that his qualities were being ignored; he was a trained specialist, he had knowledge of ten countries and ten languages, and his departmental ratings were high. Kennan argued against wasting his time on administrative jobs. In Portugal he dutifully but grumpily handled intelligence operations and negotiations for bases in the Azores.

Then, in January 1944, Kennan was sent to the European Advisory Commission (EAC) in London, where his disenchantment grew. Before he assumed the post he was admonished to give advice only when asked for it. To aggravate matters further, Kennan became ill with ulcers. The EAC was supposed to draw zones for postwar Germany, but its authority was never well defined and the military seldom explained the zonal boundaries it sought. Kennan could tolerate no more than a few months in such a spineless institution and requested a transfer. He received an unexpected boost: W. Averell Harriman, Ambassador to the Soviet Union, requested Kennan as an adviser, and Kennan was appointed counselor of the Moscow embassy. He departed in July 1944 for the land he knew best, whose leaders and foreign policy he liked least.

The years before 1944 were pivotal to the formulation of George F. Kennan's ideas about foreign policy, as well as to his self-perception as a diplomat. In his view, the Soviet Union was aggressive, made so by a fanatical ideology, historical imperatives, and ruthless leaders. The Soviets' disruptive influence in international relations stemmed largely from internal sources. Seldom did Kennan consider that the external pressure on Russia from other nations might have affected Soviet behavior. Furthermore, he held a mechanistic view of Russian foreign policy, believing it to be largely constant over time. There was little American diplomats could do but be patient in the face of Soviet officialdom's inveterate hostility toward the West. He never conceded that the West's anti-Communism, manifested by the Allied intervention in Siberia and nonrecognition, as well as the unchecked rise of Hitler's Germany and the delay in the opening of the Second Front, may have aroused some of that hostility. This intense distaste derived not only from Kennan's revulsion against conspicuously reprehensible Soviet acts but also from his own conservatism and fascination with Western civilization. His general conclusion was that the United States was too conciliatory toward Moscow.

In the period before 1944, Kennan's impatience with his own lack of power and what he believed to be his superiors' ignorance of his talents emerged. He frequently felt neglected and stymied; his ideas were rejected, and the tasks assigned him were routine and uninspiring. He was the bureaucrat who had meritoriously learned his job, but felt inhibited and restless. He was an intellectual who thought analytically, committed his ideas to paper, and tried to convey them to higher officers, where they received little attention. Kennan was a gifted man, but, as his colleagues recognized, he was also opinionated, headstrong, conceited, and cranky. He could be emotional and subjective, and his assessments of Soviet policy, although informed by uncommon knowledge, were nevertheless largely conjecture. In order to illustrate the speculative methods of Sovietologists, Charles Thayer has recorded a story about a Kentucky mountaineer asking a neighbor's boy about his parents:

"Where's yer paw?" he asked the boy.
"Gone fishin'."
"How d'ye know?"
"Had his boots on and 'tain't rainin'."
"Where's yer maw?"
"Outhouse."
"How d'ye know?"
"Went out with a Montgomery War catalogue and she can't read."
"Where's yer sister?"
"In the hayloft with the hired man."
"How d'ye know?"
"It's after mealtime and there's only one thing she'd rather do than eat."[17]

As he prepared to go again to Moscow, George F. Kennan seemed equally certain about his answers to questions relating to the Soviet Union.

Kennan arrived in Moscow on July 1, 1944, and was soon assigned the rank of Minister-Counselor; but, typically, he became irritated because his duties were largely administrative, and because Ambassador Harriman, by necessity, talked more frequently with the head of the military mission, General John R. Deane, than with Kennan. Kennan admired Harriman's efficiency, but believed the Ambassador, not a career diplomat, did not hold the Foreign Service in sufficient awe. Still deeply annoyed with the Soviet conduct of isolating foreign diplomats, feeling personally neglected, and dissenting from Roosevelt's wartime diplomacy, Kennan planned to resign.

As Soviet-American conflict over Poland heated up, however,

Harriman began to ask his Soviet expert for advice, and Kennan bombarded his chief with "protests, urgings, and appeals." Kennan, "running around with [his] head in the usual clouds of philosophic speculation," became more a political oracle and less an adminstrator, more an intellectual and less a bureaucrat.[18] From 1944 until April 1946, when Kennan returned to Washington, he drew upon the ideas he had developed in the 1930s to comment pessimistically on prospects for postwar international relations. Harriman himself was reaching for a less conciliatory policy toward Russia. Kennan's advocacy of toughness, so eloquently articulated, probably accelerated the ambassador's conversion. In January 1946, Harriman left Moscow for good, leaving a now more assertive Kennan as chargé d'affaires. Truman Administration officials read his thoughtful but alarmist cables to Washington with increasing favor.

During the 1944–46 period Kennan's writings carried several common themes. After the war, he predicted, Russia, yearning to be a great power, would build up militarily and could not be trusted to fulfill agreements. He did not believe the Soviet Union, already taxed by its severe reconstruction burden at home, would attack Western Europe, but his cables did not suggest many limitations on Soviet behavior. Indeed, he sketched a picture of the self-confident, aggressive giant, uncompromising, flushed with power, responsible for most of the world's disturbances. In early 1945 he told Harriman that Russia would seek "maximum power and minimum responsibility" in China leading to possible "domination" and "control."[19] Later that year he cabled Washington that "security and aggrandizement" constituted Soviet aims in the Near and Middle East, to be achieved through an "endless, fluid pursuit of power." Soviet tactics would include the support of nationalist groups, political intrigue, subversion, and military intervention. "One of the outstanding characteristics of Soviet foreign policy is its flexible multiformity."[20] Kennan also foresaw problems. Moscow would have difficulty arousing its citizenry to postwar sacrifice. Its postwar empire would prove unmanageable. Could the United States exploit these handicaps,? he asked.

Kennan recommended that Washington follow a policy of "political manliness" toward Russia.[21] He shared the growing sentiment in the Truman Administration that Russia should be denied postwar economic assistance until it conformed to American interpretations of treaties and proved more cooperative. Believing that Eastern Europe was lost to American influence, Kennan recommended that the United States concentrate its energies on Western Europe, building it up as a countersphere of influence, but letting no Soviet action stand without a frank rebuttal. Washington should avoid diplomatic agreements that in any way suggested American acceptance of Soviet power in Eastern Europe; but the United States should not attempt to negotiate the impossible, a rollback of Soviet influence.

Collaboration with Russia in postwar Germany, he thought, was a "pipedream." He looked upon the Potsdam Conference agreements of July 1945 with "unmitigated skepticism and despair," for though they sought to establish quadripartite control, the language was too vague. Russia could not be held to the provisions on reparations, because Moscow would take what it wanted from Germany regardless of an agreement. Dismemberment of Germany, he advised, should be accepted as a fact, and the United States should proceed independently. He feared that centralization, as provided for at Potsdam, would lead to the "communization" of Germany, because Russia would use its zone as a "springboard for a Communist offensive elsewhere in the Reich." The United States should build up economically the Western zones, because Germany was the backbone of the Western European economy. The Germans, the "strongest people in Europe," were crucial to European reconstruction.[22] A peace settlement of revenge and punishment would only arouse German nationalism and make America's task in Europe more difficult. Kennan strongly disapproved of the Morgenthau Plan, deNazification and decartelization programs, and the war crimes trials.

Kennan's ideas and recommendations in the 1944–46 period, so evident in his influential "long telegram," contain both merit and shortcomings. Russia *was* too headstrong and blunt, too suspicious and fearful. Ideology held a significant place in the Soviet outlook. Soviet propaganda attacks on the United States were crude and antagonistic. (Unlike other contemporaries, however, Kennan recognized that the USSR was usually cautious and not militarily aggressive). Soviet influence in Eastern Europe did block American influence, and there was little the United States could do to change the arrangement of power in that area. Too many agreements were vaguely worded. The Potsdam accords on Germany were probably wishful, especially given the obstructionism of the French (to whom Kennan gave scant attention). Germany did add significantly to the economy of Western Europe and, given the American perspective, had to be revived.

Yet Kennan simplified the complex, ignored contrary evidence, and indulged in a great deal of speculation. He placed too much emphasis on Soviet ideology and in so doing posited a mechanistic view of Soviet foreign policy: that it flowed not from interaction with other nations but almost solely from internal imperatives. Rarely did he suggest that actions of the non-Communist nations influenced Soviet behavior, an omission glaring in its importance. He became the psychiatrist putting the Russians on the couch. What he found was a country endlessly creating world tension; what he dismissed were the instances of compromise over Poland, Iran, the United Nations, and Germany. Nor was Communism the smooth-running monolith he depicted. Eastern Europe was not closed tight to the West before 1947 (see Chapter 3). Soviet policy there was varied, uncertain, and perhaps defensive, contrary to Kennan's description—which was not much different from Winston Churchill's account of an "iron curtain." Rips in the curtain existed. Kennan exaggerated the Soviet presence, making it appear total and much more fearsome than it really was. He cautioned the United States to abstain from meddling in Eastern Europe, but his very description encouraged thinking that affirmative action had to be taken. American diplomats attempted to use the one tool they had in the region: economic aid. They dangled loans before the Eastern Europeans, alternately extending and withdrawing them, which only aroused Soviet fears of an American intrusion and stimulated further Soviet penetration.

In these early Cold War years, Kennan seemed to hold little faith in diplomacy itself. Since the Russians could not be trusted, there was no sense in signing agreements with them.

Certainly some treaties were violated, but the issues were usually complex. Russia broke its treaty with Iran, which required withdrawal of Soviet troops in early 1946, but only because Anglo-American influence in that country bordering Russia had grown through military and economic advisers, as well as oil concessions. Kennan was one of the few American diplomats who recognized that Soviet action there was designed to head off foreign penetration, but he nevertheless seemed to accept the American advances in Iran as a basic good. In most cases, agreements were not broken, but rather interpreted differently by both sides. Too often Kennan discouraged negotiations, and opportunities for accommodation were lost. One might ask, as did the critic Henry A. Wallace, whether toughness did not simply beget more toughness.

One advocate of toughness was Secretary of the Navy James V. Forrestal; he was particularly impressed with Kennan's "long telegram." He circulated the message widely in Washington and required military officers to read it. The State Department sent it to American ambassadors, who responded with enthusiasm. Forrestal engineered Kennan's return to Washington and installed him as a lecturer at the National War College. "My reputation was made," Kennan recalled. "My voice now carried."[23] In official Washington he became a recognized philosopher of United States diplomacy; but like most philosophers, Kennan would suffer under the interpretations and emphases others gave to his ideas.

From September 1946 to May 1947, Kennan lectured at the National War College to high-ranking military officers, Foreign Service officers, and an occasional cabinet member (including Forrestal). In February, Under Secretary of State Dean Acheson asked him to participate in discussions of the Greek crisis, sparked when the British decided to withdraw troops from that country torn by civil war and Communist insurgency. After the first meeting, during which Kennan endorsed aid to Greece, he participated little. In early March he dropped in at the State Department and saw a draft of the President's message, soon to become the "Truman Doctrine" speech of March 12, 1947. Kennan complained about the sweeping language of the address and the seemingly open-ended commitments it implied.

He also opposed military aid to Turkey, because he did not believe that country stood in danger of Russian attack. His complaints came too late; the message was delivered to Congress, where it sounded a bell of panic at the same time it suggested that the United States would go anywhere on the globe to resist what it perceived to be a Communist thrust.

In late April Secretary of State George C. Marshall arrived in Washington from the frustrating Moscow Conference convinced that the Soviets had little interest in European cooperation. "The patient is sinking," he told a radio audience in reference to economic ills in Western Europe, "while the doctors deliberate."[24] Marshall moved quickly. He called Kennan to his office and ordered him to establish the Policy Planning Staff to study the European economic crisis, to make recommendations within two weeks. A harried Kennan scurried about to find personnel, and with commendable speed his new agency produced an important memorandum dated May 23. Communist activities, it began, were not at the root of Western Europe's problems but, rather, the war, which had disrupted European life. American aid, then, should be aimed not at combating Communism but at correcting the economic maladjustment "which makes European society vulnerable to exploitation by any and all totalitarian movements and which Russian Communism is now exploiting."[25] In words to appear later in the "Marshall Plan" address by the Secretary at Harvard University, Kennan's report suggested strongly that Europe itself initiate a cooperative program. Russia would not be permitted to block the new program, and the Eastern European nations would have to accept American conditions before their participation could be allowed. The conditions would be difficult for Russia to accept: endorsement of multilateral, non-discriminatory trade practices and a reduction of Soviet influence in Eastern Europe.

Kennan did not see Marshall often—it was not the general's habit to indulge in long conversations or interruptions—but the Soviet expert had unusual access to the highest officer in the State Department. Kennan also kept up his "intellectual" relationship with Secretary Forrestal, who was fascinated with a paper Kennan had presented him in January. Forrestal and the State Department gave Kennan permission to submit the paper to the editor of *Foreign Affairs*. Titled "The Sources of Soviet Conduct," the article appeared in the July issue of the journal under the name of Mr. "X." *New York Times* columnist Arthur Krock, who earlier had been shown a copy of the paper in Forrestal's office, soon dispelled the mystery of authorship, and Kennan was propelled to the center of national discussion. Commentators quickly assumed that the "X" article was government policy. Kennan was soon tagged the father of the containment doctrine; he seemed to provide the theoretical justification for the Truman Doctrine and the Marshall Plan.

The highly respected article repeated many themes expressed in the long telegram, but it emphasized ideology and left vague what the United States should do to combat Soviet expansion. Kennan wrote about the implacable Communist hostility toward the capitalist world, the doctrinaire fanaticism of Stalinist leaders, and the secretiveness, duplicity, suspiciousness, and basic unfriendliness of Soviet conduct. Russia would take its time beating down its foes abroad because Soviet leaders were confident that capitalism would collapse under its own contradictions. Moscow would accept short-run losses to ensure its long-term goal. "In these circumstances," Kennan wrote, "it is clear that the main element of any United States policy toward the Soviet Union must be that of a long-term, patient but firm and vigilant containment of Russian expansive tendencies." "Counterforce" had to be applied "at every point," "at a series of constantly shifting geographical and political points," in order to "promote tendencies which must eventually find their outlet in either the break-up or the mellowing of Soviet power."[26] Kennan pointed out that Russia had serious economic and political weaknesses. In contrast, the United States could stand as a successful model of internal stability before the world.

Although responsible for implanting the containment doctrine as the commanding principle of American foreign policy in the Cold War, the "X" article was not one of Kennan's best efforts. As he later admitted, it was too ambiguous, too imprecise. Kennan used words like "force" vaguely, and the article as a whole was cast in black-and-white, bad-guy-good-guy terms. He placed no limits, geographical or chronological, on containment, and thereby ignored the resources and will of the United

States to undertake such a long-term program. Nor did he consider the constitutional system, especially Congress's role in foreign affairs, which might place some restraints on the continuous fulfillment of containment. Kennan did not distinguish between vital and peripheral areas, and as Walter Lippmann so aptly noted, containment could become a "strategic monstrosity," with the United States supporting a host of client states.[27] Perhaps most important, Kennan never spelled out how the United States should implement containment, whether by economic, political, military, or psychological means.

Questionable was his statement that a mechanistic Soviet power "moves inexorably along a prescribed path, like a persistent toy automobile wound up and headed in a given direction, stopping only when it meets with some unanswerable force."[28] Once again he paid inadequate attention to United States power and expansion as contributing factors to Soviet behavior. Too simply, he applied one interpretive model to Russia and another to the United States: Russia's foreign policy derived from a response to internal needs not external threats; America's foreign policy derived from a response to external challenges. Elsewhere Kennan had pointed out that both ideology and traditional Russian nationalist ambitions explained Soviet behavior, but in the "X" article the latter element was underplayed. He failed to deal with the Soviet presence in Eastern Europe and the difficulties of ruling an empire of the discontented. Nor did he supply any formula for the solution of issues, thereby betraying considerable pessimism about the efficacy of diplomacy. In other government reports and lectures after the "X" essay, Kennan tried to set geographical limits for containment, always emphasizing Western Europe and Japan but including so many other areas that his attempt at precision actually smacked of globalism. In a 1948 list of priority areas worth defending, Kennan did not include Korea; but in 1950 he backed American military intervention, raising the question whether his list was so expandable that it seemed more limitless than limited. Kennan opposed policies designed to liberate Eastern Europe (because it was the one area the Soviets might risk war to maintain), and he stressed that ideology was an instrument rather than a determinant of Soviet policy. Still, what most people remembered and quoted was the strident, ill-defined language of the "X" article.

Kennan later claimed that he never meant the United States should undertake global interventionism or emphasize military means. It is frankly difficult to know what Kennan meant. He did speak frequently of the Soviet Union as a political threat and did argue that Russia would not assault Western Europe. He did oppose military aid to Greece and Turkey and the seeming permanence of American forces in Japan and Germany. He applauded the Marshall Plan as the best means to fulfill containment. He objected in 1949–50 to the decision to develop the H-bomb, questioned a strategy based upon nuclear weapons, and called consistently for tactical military units for limited wars. He doubted the need for a full-blown military alliance like NATO. On the whole, he seemed to oppose the Truman Administration's military expansion. "Remember," he lectured National War College students in October 1947, "it is not Russian military power which is threatening us, it is Russian political power." Thus, if the threat is not military, "I doubt that it can be effectively met entirely by military means."[29]

Yet, at the same time, Kennan conceded a great deal to the military point of view. Secretary Forrestal, his initial patron, was widely recognized as one of the most military-minded men in Washington. Kennan could not have been ignorant of the uses to which Forrestal was putting his ideas. Furthermore, it is a truism in international relations that to be effective politically and economically a great power must have a hefty military punch. Kennan himself had frequently spoken of an "adequate military posture."[30] The existence of military forces, he wrote in 1948, "is probably the most important single instrumentality in the conduct of U.S. foreign policy."[31] He endorsed the Rio Pact, recommended some sort of military coalition in Western Europe, toyed with the idea of American military intervention in Greece (1947), Italy (1948), and Taiwan (1949), and favored intervention in the Korean War. In the 1947–50 period he wrote *publicly* on behalf of containment without clarifying what he meant. Indeed, he reprinted the "X" article in his popular book

*American Diplomacy, 1900–1950*, published in 1951, making no alterations or comments to refine it or to indicate his differences with the military emphasis of the day.

Even if he did not intend or agree with all the means utilized by Truman to administer Cold War policies, however, Kennan seemed to offer a very broad doctrine and essentially to concede the basic notion that an aggressive and uncompromising Russia (and/or Communism) was a threat that had to be dealt with through counterforce almost everywhere and at any time. If the enemy was everywhere, as he said, then containment had to meet it everywhere; if the threat was devious and unrelenting, then the United States would have to apply whatever force was necessary to thwart it. These notions, after all, were the essence of containment. Whether he intended to or not, Kennan helped to establish undiscriminating globalism and interventionism as permanent features of American diplomacy. Kennan the intellectual would henceforth suffer as others, less reflective than himself, used his ideas to support a multitude of interventions and a military establishment of enormous proportions. The father of containment would spend much of the rest of his life attempting to disown his offspring. He felt "like one who has inadvertently loosened a large boulder from the top of the cliff and now helplessly witnesses its path of destruction in the valley below, shuddering and wincing at each successive glimpse of disaster."[32]

After the successful launching of the Marshall Plan, Kennan fixed his attention on three problems: Japan, Germany, and military defense. He believed that Japan was a strategic bastion in Asia because of its industrial might; its preservation in non-Communist hands was essential to a balance of power in that area. In early 1948, when Kennan departed for Tokyo for firsthand study, the questions of the duration of the American occupation and a peace treaty remained open. Kennan urged that the Far Eastern Commission largely be bypassed, so that the United States could act unilaterally to help Japan achieve economic rehabilitation, defense, and independence. Kennan questioned land reform, the busting of trusts, reparations shipments, and the purging of wartime Japanese from positions in government, business, and education. These programs should be curbed and a peace treaty should be delayed until Japan was able to stand on its own non-Communist feet. New directives encompassing most of Kennan's recommendations and endorsed by the President were sent to Japan in 1949, and to Kennan's surprise General Douglas MacArthur agreed. He was disappointed, however, when Secretary of State Acheson moved toward an early peace treaty, and Kennan attributed the outbreak of the Korean War to deep Soviet resentment that treaty negotiations were begun without Russia in 1950 (they were completed without Soviet signature in 1951).

Occupied Germany presented comparable problems, and Kennan recommended similar solutions. He had long opposed a punitive peace. Although American policy contained both constructive and corrective elements, Kennan wanted decartelization, dismantling, and deNazification to cease. With the coming of the Marshall Plan and the recognition that West Germany's rich deposits of coal and iron and industrial potential were vital to Western European recovery, the United States dropped most corrective measures in favor of full-scale reconstruction. In March 1949, with the Berlin Blockade still operative, Kennan took a short trip to Germany and called for an agreement with the Soviets whereby American troops would gradually withdraw after unification. Secretary Acheson vigorously opposed unification, prompting Kennan to conclude that the United States was locking itself into the endorsement of a permanent German division. Actually Kennan himself was partly responsible for this state of affairs; he had said more than once that cooperation with Russia over Germany was a "pipedream." Acheson was essentially agreeing with Kennan's earlier view.

The disagreement over Germany reflected a growing division between Kennan and Acheson. When he became Secretary of State in 1949, Acheson chose to use the Policy Planning Staff infrequently. The issue of military defense sharply widened the differences between the two men. Acheson, always seeking to work from military strength, wanted to enlarge NATO; he seemed to be preparing for a major war. Kennan did not think Russia would attack Western Europe and advocated mobile tactical units. The Secretary set in motion plans to rearm West

Germany and to develop the H-bomb. Kennan strongly opposed both because they would destroy opportunities for accords with the USSR, with which, it now seemed to Kennan, the United States could productively negotiate. He especially urged negotiations to establish international controls on nuclear weapons— and ultimately their prohibition—telling Acheson that an atomic weapon should not be looked upon simply as a conventional weapon carrying more destructive power, because the former did not "spare the unarmed and helpless non-combatant . . . as well as the combatant prepared to lay down his arms." He warned against "hypnotizing ourselves into the belief that they [nuclear weapons] may ultimately serve some positive national purpose."[33] About the same time, Kennan disputed many of NSC-68's superficial assumptions (especially that the Soviets were bent on world conquest) and recommendations for meeting the Communist threat through an accelerated militarization of containment. During the Korean War, Kennan also criticized Acheson's stretching of American objectives by sending troops across the 38th Parallel, fearful as Kennan was of exactly what happened: Chinese entry into the war.

Out of tune with the heavy military emphasis and apparent abandonment of diplomacy, and convinced that the Secretary had no confidence in his planning staff, Kennan asked in late September 1949 to be relieved of his duties. He stayed on until August 1950, when he took a position at the Princeton-based Institute for Advanced Study. As an outsider, he seemed obsessed with explaining to others through numerous lectures and publications what he had learned on the inside. He lectured Americans that they were too emotional in their foreign policy views, too easily swayed by simple answers, too susceptible to stampedes of opinion. These themes he eloquently, if too simplistically, expressed in his *American Diplomacy, 1900–1950*. In the midst of McCarthyism he protested strongly against mindless anti-Communism and defended fellow Foreign Service Officers against the Wisconsin senator's vicious attacks.

In the fall of 1951 Acheson invited Kennan to become Ambassador to Russia. Kennan was hesitant; after all, he disagreed with so much that Acheson and Truman were doing. Yet the appointment would represent a culmination of his career in

Soviet affairs and he should serve his government. So, he accepted. It did not take long for Kennan to discover that the appointment lacked meaning. He was not directed to undertake new negotiations or to reassess policy; indeed, diplomacy was discouraged. He left for Moscow "empty-handed, uninstructed, and uncertain. . . ." His disappointments multiplied. In Moscow, the embassy was bugged, Soviet agents tailed him, and the Kremlin snubbed him. Stuck in the embassy, a "gilded prison," his bitter memories of the 1930s and 1940s revived. Now he argued that much Soviet hostility was due to America's over-militarized foreign policy. He complained to Acheson that "it is not for us to assume that there are no limits to Soviet patience in the face of encirclement by American bases." When his superiors did not respond to his analysis, Kennan decided it was time to leave; but before he could, painful events intervened.

In September 1952, Kennan left Moscow for a conference in London. During a brief stop in Berlin, reporters peppered him with routine questions. One query about relations between the embassy staff and the Russian people elicited an annoyed and terribly undiplomatic reply. Being isolated in Russia, he said, was not unlike his internment at the hands of the Germans in 1941–42. The Soviet government soon declared him *persona non grata* for comparing the Soviets with the Nazis. Actually he was happy to give up the ambassadorship, although the way he gave it up violated his sense of proper conduct. The appointment of John Foster Dulles as Secretary of State reaffirmed Kennan's inclination to break permanently with the Foreign Service. Dulles wanted him out anyway; Dulles' stinging critique of containment as too defensive and his espousal of "liberation" left no doubt about his differences with Kennan. Quietly Kennan once more left Washington for the Institute at Princeton to take up his scholarly work. Articles, lectures, and books soon poured from his typewriter.

As a public figure, of course, he could not escape public issues. After Stalin's death in 1953 and an apparent thaw in the Cold War, many prominent European leaders hoped to reduce military weapons and troops in Europe. In 1957, Polish Foreign Minister Adam Rapacki advocated a "nuclear free zone" comprising Poland, Czechoslovakia, and Germany.[34] That same year, Soviet Premier Nikita Khrushchev called for a new summit meeting, and British Labour Party leaders like Hugh Gaitskill formulated plans for the removal of foreign troops from Central and Eastern Europe. The Eisenhower Administration openly rejected the Rapacki Plan, and indeed suggested that the United States might give nuclear weapons to West Germany.

A number of events stimulated these "disengagement" proposals. Khrushchev's vehement denunciation of Stalin and espousal of peaceful coexistence in 1956 suggested that a less belligerent Soviet foreign policy was developing. Indeed, the peace settlement over Austria in 1955 and the subsequent withdrawal by the United States and Russia from that country seemed encouraging. Other events, however, seemed calculated to heat up the Cold War. In 1956, the Soviets crushed the Hungarian Revolution and Britain and France attacked Egypt over the Suez Canal controversy (see Chapter 9). To some observers the failure of the Hungarian uprising indicated that the only way to relieve Soviet pressure on Eastern Europe was the removal of Soviet troops, which a mutual withdrawal of American and British troops from West Germany could accomplish. Yet West Germany had entered NATO in 1955 and continued to expand militarily. The dramatic launching of *Sputnik* in October 1957 convinced disengagement advocates that talks had to begin quickly, before America responded with an accelerated emphasis on missile development and weaponry.

George F. Kennan dramatically entered the debate in November-December 1957, when he delivered six eloquent, if vague, lectures over BBC radio (published as *Russia, the Atom and the West* in 1958). Millions in Europe heard the Reith Lectures, broadcast from London, and they were widely reprinted. The architect of containment, read the headlines, appeared to be abandoning his creation in favor of disengagement. He called for a unified, nonaligned Germany, withdrawal of foreign troops from the "heart of the Continent" (Eastern Europe and Germany), and restrictions on the deployment of atomic weapons in the region. In these lectures, unlike in the "X" article of 1947, Kennan seemed much less willing to judge Russia. "Their world is not our world." More, he intoned, "we must get over this obsession that the Russians are yearning to attack and occupy Western Europe, and that this is the principal danger. The Soviet threat . . . is a continual military and political threat, with the accent on the political." Nor should the United States meddle in the "non-European world." Developing nations should be allowed to grow in their own way; foreign aid, complained Kennan, did more harm than good. Finally, he opposed the strengthening of NATO, because in 1957–58 diplomacy rather than a "military fixation" (especially atomic) was needed.[35] Implicit in Kennan's lectures was a belief that the United States had overcommitted itself. He had not abandoned the containment of Soviet expansion; rather, he advocated disengagement as a way of liberating Eastern Europe and defusing any potential Soviet expansion. The Kennan of 1957, as a comparison of the "X" article and his Reith Lectures suggests, had been educated by changes in the Soviet-American confrontation and his realization that unlimited containment stultified diplomacy itself. The time was pregnant for tempering the Cold War, pled Kennan. The Russia of Khrushchev was not the Russia of Stalin.

A flurry of speeches, serial articles in such opinion magazines as *The New Republic* and *Foreign Affairs*, and heated rebuttals from both Truman and Acheson greeted Kennan's lectures. Opponents of disengagement, like Henry A. Kissinger, asserted that Russia remained a military threat to Western Europe. American troops, therefore, should continue in West Germany as a deterrent; NATO should not be weakened. Others protested that Soviet *political* control of Eastern Europe would not end with the removal of Soviet *military* control. The people of Eastern Europe might attempt to throw off the Communist regimes, and then, critics predicted, Soviet soldiers would storm back. The Germans themselves might not accept a reduced role in European affairs; in fact, Chancellor Konrad Adenauer vocally opposed disengagement and called for more West German military growth. Professor Hans Morgenthau, a German emigré and distinguished political scientist, was equally cautious, suggesting that a united Germany could eventually become a new and threatening German empire and that Russia might prefer a divided Germany, partially restrained by the United States, to a strong one.

Dean Acheson delivered the most stinging and unremitting critique. He charged that Kennan "has never, in my judgment, grasped the realities of power relationships, but takes a rather mystical attitude toward them."[36] The "realist" appeared to be attacking the "realist." Acheson ridiculed Kennan for his vagueness and imprecision and dismissed disengagement as "new isolationism."[37] The former Secretary of State displayed an utter lack of faith in the Soviet Union, which might, he feared, reintroduce troops in Eastern Europe, threaten Western Europe, and sign an anti-American defense pact with the new Germany. Kennan hurled a private barb at Acheson in a letter, commenting that "rarely, if ever, have I seen error so gracefully and respectably clothed." He added that "nothing . . . in Mr. Acheson's attack on the lectures would suggest that he sees the faintest differences between the Soviet government as it is today and the same government as it was eight years ago."[38]

Kennan also received enthusiastic support. British Labour party and German Social Democratic party leaders applauded him. Senator John F. Kennedy wrote Kennan in early 1958 commending his lectures for their "brilliance and stimulation" and reproving Acheson for his "personal criticisms" of Kennan.[39] Walter Lippmann was happy to point out that he had advocated disengagement as early as 1947, when he had criticized the "X" article; a decade later Kennan and Lippmann were struggling as compatriots against Cold Warriors like Acheson and Kissinger who were reliving the battles of the 1940s and who seemed to resist diplomacy in favor of military might. Kennan now trusted in Russian rationality and insisted that despite the risks, disengagement was a viable alternative to a potentially explosive Cold War confrontation. The West would not know what Russia thought without at least negotiating. In those days of rigid positions, however, opinion weighed against Kennan.

Beginning with the disengagement controversy, Kennan had intermittent contact with John F. Kennedy. While teaching at Yale University in January of 1961, Kennan received a telephone call from the new President, who offered him an ambassadorship in either Poland or Yugoslavia. Kennan welcomed the opportunity to serve again, if only to be able to leave the Foreign

Service with more honor and propriety than he had in 1952. He accepted the assignment to Belgrade and for two years cultivated contact with Yugoslav officials, gained unusual access to Tito, absorbed the varied culture of the country, and put the embassy staff to work writing a history of Yugoslavia. He noted obvious contrasts between this post and his tour of duty in Russia. Tito was not suspicious and tricky like Stalin. Foreigners were welcome in Yugoslavia, and travel was largely unrestricted. Yugoslavia practiced an independent Communism, proudly rejecting dictation from Moscow.

What bothered Kennan most was not the Yugoslavian government, but his own. In 1961, the Tito-sponsored Belgrade Conference of non-aligned nations triggered vociferous anti-Communism in the United States. Congressmen reacted sharply to some anti-American speeches at the conference and began seeking ways to punish Yugoslavia. America's ever-present, undiscriminating, and ritualistic anti-Communism flared up again. Kennan blamed the response on Eastern Europe emigré groups and the American propensity to view world affairs emotionally. Congress first disrupted the sale of planes to Tito's government. Then, in 1962, it canceled the "most favored nation" provision in trade relations between the United States and Yugoslavia—a vindictive move against a country that had few grievances against the United States and vice versa. Kennan had tried to stop this step by flying to Washington and lobbying with legislators. His failure, and his belief that Foreign Service personnel were minor figures in the American political system, "took the heart out of any belief in the possible usefulness of a diplomatic career."[40]

Earlier, in 1961, he had asked Kennedy not to endorse the "Captive Nations Resolution," which called for the liberation of peoples living in Communist countries and which Congress routinely passed each year. Kennan interpreted it to mean that Washington was committed to the overthrow of the Yugoslavian government, a policy that placed an Ambassador in an awkward position. Kennedy nonetheless submitted to political opportunism and declared a "Captive Nations Week." This action, combined with the restriction on trade, convinced Kennan that his helplessness as a representative of the United States had been publicly demonstrated to the Yugoslavs. In 1963, without bitterness and with great dignity, Kennan left the diplomatic corps. The Belgrade post was by his own estimation his most rewarding and enjoyable; however, the experience soured him further on the American political process, the intrusions of Congress in foreign affairs, and knee-jerk anti-Communism. Domestic politics, he concluded, was an "intolerable corruption" of diplomacy's "essential integrity."[41]

Yet, George F. Kennan felt compelled to go before the Senate Foreign Relations Committee in 1966 to appeal for a gradual withdrawal from Vietnam. Again he acted as the reasoned dissenter trying to persuade public opinion and the Congress— against the President—that the morass in Indochina was an unfortunate consequence of globalism. Interventionists like Dean Rusk, Walt and Eugene Rostow, and Dean Acheson had been justifying the costly expedition to Vietnam by citing the precedents of anti-Communist containment in the 1940s. Kennan refused to permit himself and his ideas to be used in this way. Under the tutelage of Senator J. William Fulbright, the Foreign Relations Committee was in an early stage of rebellion against the war and was looking for respectable allies. The recognized father of containment was willing to testify. Students and faculties had been protesting and marching, but neither the committee members nor Kennan felt comfortable with their frank and iconoclastic style. Kennan once again thrust himself into the center of a major debate by testifying that the containment doctrine was inapplicable to Indochina.

"I find myself," he remarked, "in many respects sort of a neo-isolationist." By that statement he meant that the United States should be modest and discriminating in its commitments abroad. He insisted that China was not a military aggressor in Indochina, that a military solution to what was largely an internal political conflict was fantasy, that Vietnam was an area of secondary importance to United States security, and that the war was disrupting American relations with both Japan and Russia. Facile analogies with the 1940s did not face the new realities of the 1960s. Asia was not Europe; China was not Russia. Containment should not be universal. He envisioned Vietnam as an independent Communist country like Yugoslavia; but under vigorous questioning, ambiguity crept into Kennan's presentation. He said that "if we had been able to do better in Vietnam I would have been delighted, and I would have thought that the effort was warranted. . . ."[42] This note of pragmatism seemed to undercut his appeal to principle. Yet his message was clear: The United States should curb its role as world policeman. The combination of protest marches, critical analyses by scholars, Communist victories in the hills and cities of Vietnam, and tempered dissent by public figures like George F. Kennan moved both the Lyndon B. Johnson and Richard M. Nixon Administrations, however slowly and incompletely, to a de-escalation of military intervention in Indochina.

In early 1973, a still active and alert Kennan talked frankly during a television interview. "I've been gloomy lately," he commented, because America faced massive problems at home and abroad.[43] He felt the United States should no longer try to be the dominant player on the world stage. Whether or not this was a neo-isolationist attitude (Kennan used the expression in a constructive sense), it meant limited commitments overseas, a reduction in the military establishment, more attention to domestic questions, and a frank realization that Washington had neither the duty nor the capability to establish a world balance of power. Kennan voiced skepticism about President Nixon's dramatically orchestrated trip to Beijing in 1972. Kennan was not merely suspicious of Nixon's visions of grandeur as international peacemaker; he also nurtured serious doubt about close relations with the Chinese, a consistent position for him since the 1950s. He always preferred Japan as America's "friend" in Asia and had opposed recognition of the People's Republic of China. His views toward that regime were strikingly similar to his 1940s views on the Soviet Union. In 1964, for example, he depicted China as dominated by "embittered fanatics" wedded to an ideology that encouraged the "ruthless exertion of power."[44] The Chinese were determined to destroy what Americans valued through violence and subversion, he claimed.

In the 1960s, Kennan had begun to comment frequently on domestic issues such as overpopulation and the environment. In the 1973 interview, he summarized his views toward America and its diplomacy, pointing out that the United States had no business meddling in other country's affairs until it had straightened out its own. America had become too commercial-minded, too dependent upon the automobile. A self-perpetuating technology produced unneeded inventions that "whipsawed" the American people and made them waste their natural resources. A cab ride in New York City, he said, with a cynical driver living in an unhappy, polluted city, had brought his thinking into focus. Americans, Kennan implored, needed leadership and inspiration to alleviate their serious environmental disorder. Although his prescriptions derived from a conservative perspective, they were not far removed from more radical views on the importance of preserving individualism, curtailing the machine, respecting the environment, and giving priority to domestic ills. Both recognized that the quality of American life would ultimately determine United States influence in the world. "Foreign policy, like a great many other things," he wrote in *The Cloud of Danger* (1977), "begins at home."[45]

In the 1970s and 1980s, Kennan welcomed détente, but warned that Nixon and Kissinger had oversold it, causing Americans to expect too much and then to suffer disappointment, if not anger, over the slow movement toward diplomatic accommodation. He asked for limits on American contacts abroad, pooh-poohing foreign aid to the Third World, interdependence, and multilateralism. The United States had few vital interests abroad; only Western Europe and Japan warranted an American protective shield. "The first requirement for getting on with most foreign peoples," Kennan concluded, "is to demonstrate that you are quite capable of getting on without them."[46] As for President Jimmy Carter's foreign policy, Kennan found little of merit. He criticized Carter for attempting the impossible: reform of the Soviet system through a vocal human rights policy. And, then, when Carter in 1980 became vociferously anti-Soviet and espoused the Carter Doctrine for the Persian Gulf after the Russians slashed into Afghanistan, Kennan complained that a "thundering" President was carrying a "small stick." Carter had played all of his cards at the outset, and the Russians still occupied their neighbor. "Was this really mature statesmanship on our part?," he asked.[47]

Kennan had long decried the escalation of the nuclear arms

race and America's strategic reliance on missiles tipped with weapons capable of obliterating humankind. In the 1980s, as the Ronald Reagan Administration revived images of an evil, untrustworthy Soviet Union bent on world conquest and emphasized military containment through nuclear superiority, Kennan grew more alarmed. Reagan's views on the Soviets, he wrote, were marked by "intellectual primitivism," and he deplored the "routine exaggeration of Moscow's military capabilities. . . ."[48] Kennan recommended an immediate 50 percent cut in nuclear arsenals on both sides, the de-nuclearization of much of Europe, a complete ban on testing, and a freeze on new weapons. "Cease this madness," he implored, adding his voice to the anti-nuclear movement.[49] "Morbid nuclear preoccupations" and Reagan's coolness toward negotiations with the Soviets, Kennan feared in 1983, "present a shadow greater than any that has ever before darkened the face of Western civilization." Then he summarized at length the contemporary, exaggerated American image of the Soviet threat:

It is an image of unmitigated darkness, with which we are all familiar: that of a group of men already dominating and misruling a large part of the world and motivated only by a relentless determination to bring still more peoples under their domination. By those who cultivate this image, no rational motivation is suggested for so savage and unquenchable a thirst for power. The men who suffer this thirst, one is allowed to conclude, were simply born to it—the products, presumably, of some sort of negative genetic miracle. In any case, since they were born with it and are unable to help themselves, there is no way—or so we are told—that they could be reasoned with; no basis on which they could usefully be approached; no language they could be expected to understand other than that of intimidation by superior military force. Only by the spectre of such a force—an overwhelmingly superior nuclear force, in particular—could these men be "deterred" from committing all sorts of acts of aggression. . . . Well, if this image had been applied thirty or forty years ago to the regime of Joseph Stalin it might have been nearer to reality (although even then it would have been in some respects wide of the mark). Applied to the Soviet leadership of the year 1983, it is seriously overdrawn: a caricature, rather than a reflection

of what really exists, and misleading and pernicious as a
foundation for national policy.

Russians and Americans, he went on, had to "see one another
as human beings, not as Some species of demon."[50] The
"realist" of the 1940s, retorted Kennan's detractors, had become
the dreamer of the 1970s; they quoted the "long telegram" and
"X" article back at him. The nature of the Communist threat
remained the central question in American Cold War foreign
policy.

# 8

## Beginning to Meet
## the Threat in the Third World:
## The Point Four Program

He wanted to give a memorable speech like George Washington's Farewell Address or Abraham Lincoln's Second Inaugural Address, the President told his advisers. Harry S Truman's Inaugural Address of January 20, 1949, *was* memorable—not for its cliched Cold War rhetoric of democracy vs. Communism, or for its celebration of the European Recovery Program or the new North Atlantic Treaty Organization. The United Nations, Marshall Plan, and NATO were points that people, crowded around the Capitol or watching on television, assumed the President would tout as checks against the Soviet/Communist threat. But the unexpected fourth point—that, calculated the presidential assistant Clark Clifford, would excite people bored with the familiar. "You go through the dog acts and the acrobats on up to the headline, and that's the last act of the evening. That's why it was made point 4."[1] Point Four of the Inaugural Address was stated crisply in characteristic Truman style: The United States "must embark on a bold new program" to provide technical assistance to poor peoples in the "underdeveloped areas" whose "economic life is primitive and stagnant." The President extolled self-help, the expansion of private foreign investment, and greater production to achieve "prosperity and peace."[2] Upon hearing Point Four, Secretary of State Dean Acheson,

147

sitting behind Truman on the Capitol platform, perked up; it was the first time he had heard about it. Indeed, the White House politicos had deliberately not informed the nation's chief diplomat. State Department officers wondered along with Acheson what the President was up to.

Individual Americans, businesses, federal agencies, and international organizations had long engaged in technical missions or private capital investments abroad. Technical assistance had accounted for a segment of Truman Doctrine aid to Greece, and the Institute of Inter-American Affairs had been operating for years in Latin America. What seemed new in Point Four was the frank harnessing of humanitarianism, the exportation of know-how, and overseas investment, through a single-purpose program, to the goal of coaxing nations in the Middle East, Asia, Latin America, and Africa toward American Cold War policies. By 1949 the Cold War lines had been drawn fairly tightly in Europe. But events in the developing world—what we now call the Third World—were unsettling to American Cold War managers struggling to build a stable international order governed by American power and principle. As empires deteriorated, new states emerged in the international system, and they appealed for United States economic assistance. With whom would India (freed from British colonialism in 1947) or Indonesia (independent from the Netherlands in 1949) or Libya (freed from the Italian empire in 1951) align in the Cold War? Truman officials envisioned in the popular peace and prosperity idiom, that the Point Four Program would be an inexpensive yet effective way to woo these emerging nations into the American camp while the United States continued to build an economic and military shield around Western Europe.

The leaders of the developing nations sought *development* for the future at a time when *reconstruction* from World War II claimed primary American attention. When American officials explained that the recovery of Western Europe would benefit the developing world as well, the highly sensitive leaders of the emerging states rebutted that such talk smacked of refurbished colonialism and subservience to traditional Western imperialism. When Washington pressed the new governments to take America's side in the Cold War, Prime Minister Jawaharlal

Nehru of India and others indignantly answered with non-alignment. The Point Four Program was the first of many United States efforts from the 1950s onward designed to speak directly to the needs of developing nations and to counter non-alignment or neutralism and Communism in the Cold War. If Southern Brazil or the Zambesi River Valley in Africa could be developed—if the Mesopotamia Valley could be revived as the Garden of Eden, as Truman once put it—economic growth would nurture political stability and hence a world more likely to team up with the United States against a Soviet Union, thought to be an opportunistic exploiter of human misery. As well, some strategic raw materials were in short supply in postwar America: Expanded trade with and investment in former colonial areas would reap not only profits but also enhance national security. Perhaps the hopes and expectations for the Point Four concept of technical assistance and expanded foreign investment were too high, but, having just successfully erected the Truman Doctrine, Marshall Plan, and NATO, a heady Truman could only wax optimistic about his pet project.

"I think the first problem that the State Department had to solve was what in the hell the man was talking about," recalled a State Department officer.[3] Insisting that this was *his* Inaugural Address, Truman did not want his fourth point "scooped or leaked" beforehand, so he and his staff never uttered a word about it to the diplomatic officers.[4] Point Four actually derived from an opportune merger of American domestic politics and foreign policy needs. Fresh from his stunning 1948 presidential election in November, Truman and his aides were looking for a "dramatic topic" suitable to the President's hard-earned political standing.[5] In mid-December a minor State Department official met with White House assistant George Elsey. Benjamin Hardy handed him a memorandum titled the "Use of U.S. Technological Resources as a Weapon in the Struggle with International Communism." Elsey listened with growing fascination to Hardy's earnest appeal to include technical assistance in the Inaugural Address. He adopted the idea "instantly," for "it was what I had been searching for."[6]

But it was not that easy. Hardy warned that his superiors in the State Department had rejected any mention of technical assistance in the address, because no careful plan had been worked out. The gutsy but worried Hardy, by visiting the White House, thus risked charges of disloyalty to the chief officers of his department. Presidential assistants had squabbled before with the State Department over the substance of speeches; in this case they were not about to reject a new foreign policy program which seemed to carry domestic political advantage just because the foreign affairs bureaucracy did not have plans ready. One month before the address, then, Elsey had a "speech in search of an idea, and an idea in search of a speech."[7] He and Clifford talked to a receptive President, and Point Four, to the State Department's immediate distaste, became a new if shapeless instrument in America's foreign policy arsenal.

Point Four soon became victimized by bureaucratic warfare between Pennsylvania Avenue and Foggy Bottom, State Department lethargy, events which detracted from the program's importance, and congressional inertia. The "bold new program" began as a catchy slogan, not as a plan with an administrative structure or priorities and procedures for implementation. The White House put the State Department in the embarrassing position of having to create a program hastily to capitalize on the dramatic and popular effect of the address. Annoyed State Department officials, although basically friendly to the technical assistance concept, were bewildered by the large gap between rhetorical expectations and implementation. Truman admitted on January 27, 1949, that Point Four was not yet a "program": "I can't tell you just what is going to take place, where it is going to take place, or how it is going to take place. I know what I want to do."[8]

The first State Department report to the President in March 1949 disappointed White House advisers. One complained that the diplomatists had "no idea of what we are actually going to do," leaving the Truman Administration open to the charge that the Point Four program, "after more than two months of labor, had bogged down in a mire of words." The presidential assistant David Lloyd sought a more "concrete" statement, with emphasis on areas where Communist pressure seemed strongest—the Middle East, India, and Southeast Asia.[9] The lethargy nonetheless persisted. An official in the Council of Economic

Advisers reported in mid-1949 that the Secretary of State "has been far too busy with the problems of Germany, the North Atlantic Pact, etc., to give Point Four much personal attention." Nor had public opinion been marshaled.[10] The foot-dragging derived in part from a ponderous debate within the administration over whether the Point Four program should be run by the State Department or a new independent agency headed by a corporate executive. The State Department won. Clifford recalled that Point Four ultimately became side-tracked by the bureaucracy. "We weren't able to get the people into it that could have given it life and color and drama; it just became another State Department program, and finally, I think, pretty well died on the vine."[11]

Not until June 24, 1949, did the President ask Congress for Point Four legislation, when he reminded the senators and representatives that hungry people might "turn to false doctrines" unless they received help. Truman recommended a modest first-year appropriation of $45 million.[12] His request came so late that Congress failed to consider it before adjourning at the end of the year, although some hearings were held. By December 1949 Lloyd was blasting the State Department for "deferring" Point four as a "major effort."[13] As late as May of the next year, about the time Congress finally acted, Ambassador to the United Nations Warren Austin concluded that "Point IV ideas are under 'wraps.'"[14]

After rough treatment in the Senate, with Republicans voting overwhelmingly against, Point Four passed in May 1950 as the Act for International Development. The President signed the bill on June 5, less than three weeks before the outbreak of the Korean War, and eighteen months after he had first proposed the new program. The act authorized only $35 million for technical assistance and insisted that recipient nations provide a healthy investment environment for foreign capital. Congress caused further delay by waiting until September to appropriate only $26,900,000 for Point Four—an insignificant sum compared to $2.25 billion for the Marshall Plan and $342,450,000 for occupied Germany at the same time, although a smaller figure would be expected for technical assistance because it did not involve large public capital grants. Not until early December

1950 did the new Technical Cooperation Administration (TCA) within the State Department get a director.

During these long months of unenthusiastic activity within the executive bureaucracy and Congress, a public debate on the merits of the Point Four got under way. Truman had early worried about opposition. He advised Elsey: "Just don't play into the hands of crackpots at home—no milk for Hottentots." The "natives" would have to help themselves.[15] Having often tapped the congressional purses for foreign aid, the Truman Administration was reluctant to ask for "bold new" appropriations. Truman popularized Point Four as a practical, inexpensive, common-sense program that would both fight Communism and insure American prosperity. But some skeptics asked why another program was needed when existing programs could simply be extended to handle technical assistance. Others protested against greater spending on foreign aid after large appropriations for the Marshall Plan, the Export-Import Bank, and the World Bank. Penurious Republicans like Senator Robert A. Taft of Ohio recalled the New Deal to dismiss Point Four as another wasteful W.P.A.—this time global.

Partisanship in the Senate and House was conspicuous, perhaps because Truman articulated Point Four as *his* program. Also, Democratic leaders were under fire for increasing their party's membership on the Senate Foreign Relations Committee from a 7-6 to an 8-5 ratio. Republicans had not been pleased either in early 1949 with Truman's appointment of Dean Acheson as Secretary of State. Heatedly partisan congressional campaigns blossomed in late spring 1950, with the "Democratic" Point Four an occasional Republican target. The collapse of Nationalist China and Mao Zedong's triumph also undercut the administration's credibility and bargaining power, and the Korean War cast further gloom. Bipartisanship had eroded. Into this unfriendly arena entered Point Four.

As Point Four languished for months, it became more obvious that overseas American private investment would become central to the successful operation of the new program. Truman had not highlighted that role in the Inaugural Address, placing more emphasis on technical assistance, but in his June 24, 1949, message to Congress, the President devoted considerable atten-

tion to the encouragement of private American capital investment abroad. And in his State of the Union message in January
of 1950, he re-emphasized the need for large amounts of capital.
For American businessmen, to whom Truman appealed for
help, "foreign investment—not technical assistance—is really
the crux of Point Four," argued a Ford Motor executive.[16]

American businessmen considered the postwar foreign investment climate uninviting. Nationalization threatened American property from Western Europe, where the French protested
against Coca-Cola bottling plants ("cocacolonization"), to developing nations, where growing nationalistic and revolutionary
stirrings frightened investors. American businessmen shied
away from some areas, too, because of Communist propaganda,
foreign taxation, social legislation which detrimentally affected
profits, favoritism to nationals, inadequate dollar exchange
procedures, and increased governmental interference. American private capital, then, tended to bypass high-hazard regions
in favor of Western Europe or quick-profit areas rich in raw
materials such as oil.

The Truman Administration hoped to reduce these impedi-
ments through "investment guarantees." Recipients of Point
Four technical assistance were required to create attractive
private investment conditions, but Washington knew that stable
conditions would develop slowly. Thus Truman asked Congress
to guarantee, through the Export-Import Bank, American investments. The administration appealed directly to businessmen at special Washington conferences, recognizing, as an
Under Secretary of Commerce put it, that the "success of this
program depends on the degree of confidence which the business community has in it."[17] Businessmen, as did labor and civic
groups, endorsed Point Four, but there was vigorous disagreement on the merits of the investment guarantees. The National
Foreign Trade Council, representing large international corporations, complained that investment guarantees (guaranteed by
Washington) shifted the responsibility for protecting investments from recipient nations, where it should be lodged, to the
United States. Many investors preferred bilateral protective
treaties, which the Truman Administration was then signing, to
guarantees. When the Act for International Development

passed in 1950, the provision for guarantees was absent, although it would be added the following year.

Point Four, with or without investment guarantees and protective treaties, stimulated little American business activity in high-risk, low-profit areas. "What capital wanted most, apparently," *Fortune* magazine concluded, "was a healthy investment climate, not an insurance policy against contracting malaria."[18] Investment hazards persisted. In 1954, Clarence B. Randall, head of a presidential commission studying American economic foreign policy, found that only Turkey, Greece, and Panama had altered their laws enough to suit American capital. Between 1950 and 1957, American private investment abroad increased impressively from $11.8 billion to $25.2 billion. Yet this significant growth took place largely in traditional nations already part of the American sphere of influence. Five billion dollars, or nearly half of this increase, flowed to neighboring Canada, and $2.4 billion went to Europe. Latin America realized an increase of $3.5 billion. But developing nations in Asia, Africa, and the Middle East enjoyed comparatively little investment. The Middle East and Africa received new American investments amounting to only $800,000,000 between 1950 and 1957. By 1957 American investments in Asia and the Middle East totaled $2 billion; but $1.5 billion of that figure was invested in petroleum. Hungry India in the 1950–57 period saw American investments increase from $38 million to only $113 million. The most critical sector of the heavily populated and poor countries in Asia and the Middle East—agriculture—accounted for private American investments of only $36,000,000 in 1957, or an increase of a mere $1 million since 1950.

If the private investment component of Point Four proved inadequate, so ultimately did the technical assistance program. Skeptics sneered that Point Four was neither very bold nor very new. Appropriations for technical assistance did grow; Congress provided $147,900,000 in 1952 and $155,600,000 in 1953 for the Technical Cooperation Administration which implemented Point Four. But in mid-1953, when the TCA was submerged in the military-minded Mutual Security Agency, Point Four lost its identity and specific appropriations. Indeed, the Point Four program soon became a link in America's defense system with the special function of increasing the shipment of strategic raw materials to the United States during the Korean War. The Truman Administration increasingly understated the altruistic aspects of Point Four to argue that the program was an inexpensive bargain for the American taxpayer. As one pamphlet stressed in early 1953, the cost of Point Four for that fiscal year would be "less than the price of one battleship."[19]

Disappointed supporters of Point Four chastised Truman for creating a program with the "head of a lion and the body of a mouse."[20] The State Department at times seemed more concerned with fighting off attempts to shift the TCA away from its jurisdiction than with energizing Point Four. Personnel in North Africa grumbled about the non-delivery of equipment long promised. Supreme Court Justice William O. Douglas complained that technical assistance shored up "corrupt and reactionary regimes" and "feudal systems" which subverted necessary social and economic reforms.[21] But the United States seemed to be caught in a dilemma. To improve food production, Washington gave assistance to those few people who owned the land and who at the same time resisted land reform. If Washington pressed the landed elite to reform, it invited cries of "imperialism." If America did not demand corrective changes, it was charged with reactionary tendencies. Foreign aid to developing nations had its limits, and when it came to the question of securing Cold War partners, American leaders looked less at the form of foreign governments or their attitude toward social reform and more at the degree of their allegiance to American foreign policy, as was the case with the Philippines.

Far from the maneuverings in Washington, Point Four technicians circled the globe. Numbering 1,500 in 1953, they visited thirty-five countries. Digging compost pits in India, administering smallpox vaccine in Liberia, combating a locust plague in Iran, conducting a water survey in Saudi Arabia, instructing teachers in the Peruvian Andes, or eradicating malaria in Venezuela, American technical experts ameliorated life for people in isolated districts. Then, too, foreign nationals—one thousand in 1953—traveled to the United States for study. Yet the appropriations, the individual efforts, and the fanfare for Point Four never matched the magnitude of the problem. Point

Four came to mean expanded raw materials imports to the United States, but comparatively little to the destitute abroad.

Foreign critics growled their complaints too. The Indian delegation to the United Nations General Assembly protested that technical assistance was designed to ship raw materials to industrialized nations. By 1953 Syria had refused direct American technical help, and Egypt's Gamal Abdel Nasser labeled Point Four an instrument for colonial penetration. Many Arab states protested American aid to Israel by snubbing Point Four. Colonialism also impeded the program: Portugal in 1952 would not let American technical experts into Angola and Mozambique, and Belgium rejected Point Four for the Congo. Another problem, critics noted, was not only that American assistance benefited elites in some countries, but also that the United States sometimes promoted projects more helpful to itself than to the developing countries. Washington, after all, wanted political and strategic benefits from Point Four. It was a Cold War program to meet the Communist threat. The recipients, on the other hand, looked for national economic gains. Airfields and highways in Afghanistan, Jordan, Saudi Arabia, Thailand, and Vietnam satisfied American military purposes and American business, but probably served minimally the economic development of those nations. This record, coupled with private American business' primary interest in profit maximization rather than general social and economic improvement (investments in oil rather than food production, for example), blemished Point Four and actually undercut its intended usefulness as a way to win friends, especially among nonaligned, developing countries. The problem with Point Four, to argue with Clark Clifford, was not simply bureaucratic entanglement and neglect but rather the assumptions and self-interest of the program which clashed with the goals and needs Third World leaders identified for their impoverished nations.

Critical of Point Four as a self-interested American program, some developing nations turned to the United Nations. But, as in the cases of the Truman Doctrine and the Marshall Plan, in Point Four Washington chose to control funds directly and largely sidestepped that international body. President Truman said that Point Four would cooperate with the United Nations whenever practicable, yet it was in fact his fear that the State Department would place too much emphasis on the role of the international association in technical assistance. Although the Truman Administration rejected a major program under the auspices of the United Nations, it did grant $12.5 million to small United Nations technical assistance projects from June 1950 to December 1951, and about $15 million each year thereafter in the 1950s.

Hastily announced, originally neglected by the State Department, lost in the turmoil of more dramatic Cold War issues, hampered by limited congressional appropriations, resisted in many parts of the Third World, spurned by American businessmen, and ultimately diverted to military purposes and strategic materials stockpiling, Point Four faltered early. "Economic aid to the developing world," President Dwight D. Eisenhower lamented in 1955, "has the political appeal of an ordinary clod in the field."[22] Point Four exaggerated the changes that could be realized from technical assistance, and thus failed in its goal of aligning the Third World with United States foreign policy and ameliorating the economic conditions that bred social disorder and political instability.

In the wake of Point Four's meager performance and with the rise of American fears of a Soviet "economic offensive," whose new if small aid projects threatened to accelerate the thrust of "international Communism" into the Third World in the late 1950s, President Eisenhower and his advisers reconsidered American economic development policy. Eisenhower knew well that people in the developing world sought "freedom from grinding poverty," and "no nation, however old or great, escapes this tempest of change and turmoil"—the "flames of conflict" that flare up from "desperate want."[23] The question simply became one of "aid or chaos," remarked the President, and he elected to improve and expand aid.[24]

For years the United States had blocked attempts within the United Nations to form an agency to extend loans on lenient terms to developing nations—the Special United Nations Fund for Economic Development (SUNFED). But intensified Cold War competition moved the Eisenhower Administration in 1957 to agree finally to endorse a scaled down SUNFED called the

"Special Projects Fund." The United States that year also created the Development Loan Fund (DLF). With an initial annual appropriation of $350 million, the DLF began to make low-interest loans, repayable in local currencies, to countries like India. In the next few years, the Eisenhower Administration also established the Inter-American Development Bank, beefed up the Export-Import Bank, expanded the "PL 480" program of food assistance, and began to energize the World Bank to make loans to developing nations. If Point Four had failed to provide an adequate instrument to win the Third World's favor, it at least supplied the Cold War argument and lofty rhetoric of human uplift and democracy that echoed in the speeches and programs of Presidents Eisenhower and John F. Kennedy. But both leaders exaggerated the Communist threat to the Third World and claimed too much for the remedial effects of foreign aid, as Eisenhower soon discovered in the turbulent Middle East.

# 9

## Threat to the Middle East?
## The Eisenhower Doctrine

On an evening in November 1954 CIA agent Miles Copeland, undercover as an American company representative, drove Colonel H. Alan Gerhardt and Major Wilbur C. Eveland to a villa in suburban Cairo. Dressed in civilian garb for their secret mission, the two American officers entered through the back door. Moments later the tall, smiling leader of Egypt, Gamal Abdel Nasser, bounded in. They exchanged pleasantries. Soon seated at the dining room table, Nasser produced two packs of Kent cigarettes and proceeded to chain-smoke them while Military Chief of Staff Abdel Hakim Amer handed the visiting Americans a substantial request for military items—tanks, heavy artillery, bombers, and more—worth at least $100 million. They discussed the terms of an arms agreement. Balking at the fine-print provisions, the English-speaking Nasser quipped that "half the secretary of defense's office must be lawyers."[1]

Soon Gerhardt began to sketch the virtues of Egypt's participation in a regional alliance for the Middle East similar to NATO and the new Southeast Asia Treaty Organization (SEATO). Amer and Nasser became excited. "A regional arrangement might serve *your* purpose," said Amer, "but before we can say what serves our purpose we've got to know whom we're going to be fighting. Whom *are* we going to be fighting?"[2] Russia, of course, replied the Americans; but perhaps some language could be devised that did not name a specific enemy. The always frank Nasser protested against such theoretical talk. After all, he insisted, if Egypt and other Arab states warred again with America's friend Israel, surely the United States would withhold military aid from the Arabs. So, specifying the threat mattered. And, although Secretary of State John Foster Dulles had pressed the Arabs to realize that "international Communism" constituted the primary danger in the Middle East, Nasser said that he had "seen no signs of Russian hostility except to the defense organizations" that the United States was "erecting to surround the Soviet Union."[3] No, there were only two enemies in the Middle East: the Israelis and the British. "The Arabs don't know anything about the Russians. It is foolish to try to stir them up to a fear of Soviet invasion."[4]

Egypt and the United States did not sign a military agreement. Soon Nasser vigorously opposed American efforts to build a regional alliance; Soviet arms flowed to Egypt; the Israelis and Arabs renewed their warfare; and the United States, fearing that the Soviets would exploit a vacuum opened by the decline of British and French influence in the region after the Suez crisis, declared a new version of containment, the Eisenhower Doctrine. The Marine invasion of Lebanon in 1958 became the most conspicuous sign of the failure of American Middle Eastern policy—failure derived from exaggerating the Soviet threat in the volatile area.

To official Washington the Soviets had been threatening the Middle East since the end of the Second World War. Believing the Soviet menace to be global, Americans assumed that Kremlin leaders, like the czars before them, coveted lands to the south. Confrontation in Iran in 1946, Truman Doctrine aid to Greece and Turkey the following year, and clandestine American involvement in a Syrian election (1947) and coup (1949) demonstrated increasing United States activity in the region. With the Cold War in apparent stalemate in Europe in the 1950s, with new, poor, unstable states emerging from colonialism into nationhood in the developing world, and with the United States busy committing itself elsewhere—in Indochina, Central America, and Europe—the Middle East seemed vulnerable. "You could practically hear the Russian boots clumping down over those desert sands," remembered Raymond Hare, American ambassador to Egypt, 1956–59.[5] Major General George H. Olmsted also envisioned a Soviet onslaught. He stuck flag-pins into a map to illustrate America's overseas defensive posture. Noting the absence of pins in the Middle East, the officer warned that the region constituted a defense "vacuum."[6]

Americans sought to fashion a new order—to merge the Middle East into their global strategic system, adding a Mideast defense organization to NATO and SEATO so that the Soviet Union would be ringed with American allies and bases. American strategists valued the area's petroleum that fueled Western European economies and armies. Of Europe's crude oil imports in 1955, 89 percent came from the Middle East. "The West must, for self-preservation, retain access to Mid-East oil," remarked President Dwight D. Eisenhower.[7] An expanded American military position in the region would insure the flow of petroleum to NATO allies, as well as deny Soviet access to the strategic commodity. American leaders eyed more airbases like that at Dhahran, which Saudi Arabia provided in 1951. American officials did not fear an immediate Soviet military attack; rather, they anticipated that, in the event of a general war, the Soviets would attempt to seize the Middle East, especially the Suez Canal and pipelines, to deny the West its vital oil. In a general war, too, the United States expected to utilize Mideast bases to attack the USSR, presumably, under the doctrine of massive retaliation, with atomic weapons.

But the threat to American objectives and interests in the Middle East in the 1950s, argued American leaders, was not merely military or directly Soviet. They also identified Arab nationalism and neutralism as dangers. Feeding on anti-Western passions, fervent Arab nationalism inflamed tensions with Israel and nurtured radical politics against traditional rulers. In this environment of instability, Americans feared the Arabs might interrupt oil shipments, or the Soviets might exploit opportunities to diminish America's strategic-economic interests in the region. Nasser's leadership of the non-aligned or neutralist movement, evident in his prominence at the 1955 Bandung Conference of Third World states, also unsettled

Washington. Neutralism, thought Secretary Dulles in the "if you're not with us, you're against us" mode, played into the hands of the Communists. Dulles did not confuse nationalism or neutralism with Communism, as some of his critics charged. No, he distinguished between them—and loathed both for serving Soviet purposes. Nasser, on the other hand, repeatedly argued that nationalism provided the best defense against Communism. He claimed, too, that Americans exaggerated the Soviet threat, and by attempting to build anti-Soviet barriers they actually stimulated greater Soviet activity in the Middle East.

Besides seeking to thwart Communism, Arab nationalism, and neutralism and to protect and expand American economic-strategic interests, the United States sought to sustain the Jewish state of Israel, which, since its stormy founding in 1948, with America's endorsement, had developed a special relationship with the United States. Israel was the strongest military force in the region and constantly at war with Arab neighbors hell-bent on destroying the new nation. As long as the Arabs preoccupied themselves with Israel, Dulles concluded, they would not pay adequate heed to the menace of Soviet Communism. Thus Dulles worked to reduce Arab-Israel antagonisms, but to no avail. Israel itself often spoiled American efforts at mediation. In 1953, against both United States and United Nations protests, Israel moved its government offices to the city of Jerusalem, a disputed territory, and attempted to divert the waters of the Jordan River to its own uses. Shortly after President Eisenhower's special emissary Eric Johnston announced his cooperative, regional plan to develop the Jordan River Valley to expand arable land through irrigation, and thus facilitate the resettlement of hundreds of thousands of Palestinian refugees, Israeli commandos staged a "retaliatory" attack against Qibya, Jordan. In the "Lavon affair" of 1954, Israeli agents tried to obstruct improving Egyptian-American relations by burning American buildings in Cairo and then blaming the terrorism on Arabs. That year, too, Israel quietly bought arms from France in apparent violation of the 1950 Tripartite Declaration of United States, Britain, and France; that agreement provided for limits on weapons sales to the Middle East to forestall an arms race.

To cite other examples, in early 1955, a few days after the United States-backed defense organization called the Baghdad Pact was launched despite Israeli protest, Israel's forces attacked Egyptian soldiers in the Gaza, killing thirty-seven. Although Egypt had actually tried to curb Moslem terrorists who used Gaza to harass Israel, the Israelis may have conducted the "reprisal raid" to expose Egypt's military weakness and to delay American efforts to persuade Israel to make a compromise peace. As Israel's banker, the United States became implicated in the Arab mind as a co-conspirator. The United Nations Armistice Commission and Security Council condemned the raid. But, more important, Egypt vowed at that point to seek military assistance from the Soviet Union. At another time, when Dulles recommended that the United States loan funds to Israel so that homeless Palestinians could receive an Israeli "compensation which is due," Israel torpedoed the proposal.[8] Israel's attack against Egypt in the 1956 Suez crisis, of course, drew stern protests from Washington. Even before that, Assistant Secretary of State Henry Byroade frankly advised the Israelis that they should "drop the attitude of a conqueror and the conviction that force and a policy of retaliatory killings is the only policy that your neighbors will understand."[9] Dulles told the President that the United States would have to find ways to "dissuade the Israelis" from taking "precipitate steps which might bring about hostilities and thus endanger the whole Western position in the Near East to the direct advantage of the Soviets."[10] Israeli behavior—so resistant to American cautionary advice—when combined with intense Arab hostility toward both Israel and Britain, repeatedly frustrated American plans to convert the Middle East into an anti-Soviet bastion.

Although they attempted in the early 1950s to avoid a distinctly pro-Israel posture, American decision-makers did not lean heavily against Israeli transgressions, first, because Israel, however uncooperative, remained a solid, militarily strong ally, and, second, because they felt hampered by the Jewish community in the United States, always politically active in defending

Israel. "We are in the present jam because the past administration had always dealt with the area from a political standpoint and had tried to meet the wishes of the Zionists in this country and that had created a basic antagonism with the Arabs," Dulles too simply complained. "That was what the Russians were now capitalizing on."[11]

To meet the perceived Soviet threat in the Middle East, the United States exercised several instruments. By 1953 American companies produced about 70 percent of the region's oil. The Arabian-American Oil Company (Aramco) dominated Saudi oil fields and operated as a subsidiary of Standard Oil of California, Texaco, Socony, and Jersey Standard. J. Paul Getty held a profitable concession in the neutral zone between Saudi Arabia and Kuwait; Jersey Standard and Socony operated in Iraq; Texaco and Standard Oil of California pumped in Bahrain; and Gulf Oil was active in Kuwait. After Iran's nationalist government was toppled in 1953, with major assistance from the CIA, American interests gained a stake in that nation's oil industry. Eisenhower Administration officials counted on the American companies to keep the oil flowing. They did not prosecute them under the anti-trust laws even when they entered into cartel agreements, for to diminish them at home was to weaken the companies' valued position in the Middle East.

The CIA also served as an instrument of American foreign policy. In Egypt, Miles Copeland worked undercover as a representative of the management-consulting firm of Booz, Allen, Hamilton International, and another CIA agent, Frank Kearns, did so as a CBS news correspondent. In 1952 the CIA tried to persuade King Farouk to moderate his policies so that he could hold power against a building revolution. After Nasser's Society of Free Officers succeeded against the monarch, the CIA helped that new government start a broadcasting station, "Voice of the Arabs" or Radio Cairo—a project Nassar eventually turned toward anti-American propaganda. In Cairo, the CIA circulated anti-Moslem books and tried to associate the distribution with the Soviet embassy. When the CIA passed $2 million to Nasser, he indignantly took what he considered clumsy bribe money and defiantly built a tower in Cairo popularly known as "Roosevelt's erection" after CIA agent

Kermit Roosevelt. In Lebanon CIA agents masqueraded as oil company employees; in Syria they plotted with factions to overthrow a pro-Nasser government; in Iran they trained and supplied the secret police, SAVAK. The CIA financed Air Jordan and Iranian Airways. Under the code name NOBEEF, the CIA subsidized King Hussein of Jordan.

The United States also extended economic and military assistance to advance its foreign policy aims. Point Four technical assistance spurred the construction of schools, funded irrigation projects, and improved agricultural techniques. Johnston's abortive Jordan River project and millions of dollars for the United Nations Work and Relief Agency—dedicated to helping Palestinian refugees—were also designed in the peace and prosperity idiom to improve economic conditions and in turn produce political order. As for economic assistance, in the period 1952–56, Israel received $256.6 million, Egypt $62.2 million, and Jordan $25.2 million; Lebanon and Iraq received a few million; but the largest aid program was in Iran, which enjoyed aid of $271.2 million. The Export-Import Bank was also active in the Middle East; it granted loans to Israel (for 1945–55, $135 million), Egypt ($7.3 million), and other Arab states ($47.3 million).

Military aid and missions went to Iran, Saudi Arabia, and Iraq, and in 1950–55 American arms were shipped to Egypt ($1.3 million), Syria ($418,000), Saudi Arabia ($20.5 million), Lebanon ($169,000), Jordan ($3,740), Iraq ($157,000), and Israel ($7.8 million)—or a total of $30.4 million. But the greatest American military activity was measured in alliance-building. The Truman Administration had tried but failed to launch a Middle East Defense Organization. Many Arabs were wary of Western-initiated pacts, for their experience with Western imperialists left sour memories, and the British still occupied Egypt—as they had since 1882. Secretary Dulles was determined to break the deadlock and fill the Middle East "vacuum" in the American defensive network. He traveled to the region in May 1953 and there encountered the formidable Nasser. When Dulles raised the specter of the Soviet menace, Nasser answered: "The Soviet Union is more than a thousand miles away and we've never had any trouble with them. They have never attacked us. They have never occupied our territory. They have never had a base here, but the British have been here for seventy years."[12] Dulles drew two conclusions: first, that his plans for a regional alliance with Egypt as the "cornerstone" could not proceed until the British departed that angry nationalistic country, for only then would the Egyptians recognize that the Soviet Union, not Great Britain, was the enemy; second, that the place to begin constructing a regional alliance lay in the "northern tier" nations of Turkey, Pakistan, Iran, and Iraq, not Egypt. It could be wooed later with military and economic aid, including funds for building the Aswan Dam, as well as with American help in easing the British out (accomplished in part by the Anglo-Egyptian treaty of July 1954).[13]

The United States' first successes came in April 1954, when Turkey and Pakistan signed a military accord and the United States and Iraq initialed an agreement for American military assistance. When Pakistan later in the year joined SEATO, SEATO and NATO in essence became intertwined through Turkey's membership in the latter. Then, in February 1955 Iraq and Turkey formed the Baghdad Pact, thereby linking NATO, SEATO, and the Middle East in a chain of anti-Soviet military agreements. All the while the United States kept pressing Egypt. "This hammering, hammering, hammering for pacts will only keep alive the old suspicions" of Western imperialism, Nasser protested.[14] The Communists, he warned, would exploit those suspicions. Nationalism, not Western intrusions, constituted the best restraint upon Communism. In March, Nasser met the Baghdad Pact with an Egypt-Saudi Arabia-Syria pact. Soon Britain, Pakistan, and Iran joined the Baghdad system (which in 1959 changed its name to the Central Treaty Organization or CENTO). A war of pacts was not what American officials either anticipated or welcomed. They had mixed feelings toward the Baghdad Pact itself, even though it basically fit Dulles' "Northern tier" scheme. The British became very active in its formation and administration, and Dulles worried that the continued British presence in the Middle East would exacerbate Arab nationalism and derail the American drive to bring Egypt into a regional security agreement. Dulles was hesitant too, because, should the United States decide to join, membership would require a two-thirds vote in the United States Senate, where Jewish lobbyists might very well engineer a rejection. Moreover, because Iraq and Egypt were rivals, Eisenhower and Dulles opposed Iraqi membership in Baghdad; but the British insisted. Then, as Washington feared, Nasser denounced the Baghdad Pact not only as a menacing Western device but also as a scheme to split the Arabs and isolate Egypt. Although the Eisenhower Administration was growing impatient with Egypt, in early 1955 it still hoped for Egyptian concessions to a defense pact and peace with Israel.

Meanwhile, American-Egyptian military talks faltered. The United States set terms the Egyptians would not accept: payment in hard currency, which the Egyptians found scarce, and American supervision of the aid through the stationing of American military personnel in the country, which reminded sensitive nationalists of haughty British military missions. Nasser lectured Americans that they should understand the "psychology of the area," because "words like 'joint command,' 'joint pact,' and 'mission' are not beloved in our country because we have suffered from them. . . ."[15] He said repeatedly, if American aid did not materialize, he would look to Moscow for help on better terms. American leaders thought he was bluffing. He was not. The founding of the Baghdad Pact, the Israeli attack on Gaza, and lack of movement on American aid coincided to persuade him to seek Soviet assistance.

In September 1955, American officials learned that the Egyptians had just struck an arms deal with the Czechs, who were standing in for the Soviets. In exchange for Egyptian cotton, Soviet weapons would soon be on their way to Nasser's forces. Washington hurriedly offered arms on credit if the Egyptian leader would scuttle the Soviet agreement. Too late, replied a buoyant Nasser, proud of his statement of independence from the West and now celebrated throughout the Arab world for his accomplishment. The Egyptian ambassador to the United States, in Cairo at the time, was not pleased, for he feared American retaliation. "Guatemala," he kept saying to Nasser. Ahmed Hussein's reference to the 1954 CIA coup in Central America did not sit well with Nasser. "To hell with Guatemala," he snapped; he would risk American covert subversion in order to obtain the jet fighters and twin-engine bombers his military needed for defense against an anticipated Israeli attack.[16]

Washington officials thereafter cooled toward Nasser and warmed to the Baghdad Pact. "We can strangle him if we want to," asserted Dulles.[17] The United States sent an observer to the alliance's first meeting, assured the signatories of American cooperation, expanded military aid to members, and joined some of the alliance's committees. (Washington also supported the sale of aircraft from Canada and France to Israel.) The United States was "in the pact but not of it, a participant for practical purposes but without legal commitments."[18] The American relationship with the Baghdad Pact pleased no one. The Israelis roared against strengthening any Arab state. Arabs and non-Arabs in the pact complained against America's less than full support. The Saudis, Jordanians, and Syrians refused to join. The Egyptians blamed the United States for exacerbating intra-Arab antagonisms and reviving British imperialism. The British, still competing with Americans in the area, criticized Eisenhower officials' failure "to put its weight behind its friends, in the hope of being popular with their foes."[19] The Baghdad Pact proved a large mistake; rather than blocking Soviet influence in the Middle East, it gave Egypt and its friends good reason to invite the Russians. By 1956, then, the Eisenhower Administration was tangled in Arab-Israeli, Arab-Arab, and Arab-British rivalries, while the Soviets, through their stunning weapons deal with Cairo, leaped over the northern tier into the Arab core.

What exactly were the Soviets up to in the Middle East before that agreement and before the Suez crisis? Actually never as much as American rhetoric had it; nor was Soviet activity in the region of such a military nature as to have justified America's drive for military containment. Despite American alarms about Russian boots pounding desert sands, the Soviet presence was minimal and likely to remain so because of local conditions: strong anti-Communism, fervent devotion to Islamic religion, preference for neutralism, pan-Arabism, and wariness of meddlesome outside powers. "The Communists are at most an internal danger to Egypt," Nasser informed Dulles, "but one that can be dealt with by social reform."[20] As for the possibility of Soviet domination, Nasser reminded Ambassador Hare that "we haven't gotten rid of one imperialism to take on another one."[21]

The Soviets, of course, were not mere bystanders in the Middle East. Geographical proximity, long-standing competition with the British in Iran, inheritance of Russian desire for Mediterranean influence, and hostility to American alliance-building at its borders compelled studied attention. Before the Second World War the Soviets were barely to be seen in the Middle East. After the war, Moscow supported autonomy for Azerbaijani rebels in northern Iran and pressed Turkey for shared power over the Dardanelles, but by 1947 it had backed away from confrontations in both places. The Kremlin favored the partition of Palestine to speed the British exit and rapidly recognized the new state of Israel. When Nasser's nationalists took power in Egypt in 1952, the Soviets dismissed them as reformists, bourgeois "stooges of western imperialism," and urged the nation's workers to challenge the new government.[22] Soviet leaders also denounced the Anglo-Egyptian agreement of 1954 and Nasser's outlawing of the Communist Party. Arab Muslims in turn did not appreciate the Soviet mistreatment of the Muslim community in Russia.

In late 1954 and early 1955, in apparent reaction to the CIA-assisted coup in Iran, Turkish-Pakistani pact, and Baghdad alliance, Soviet policymakers shifted tactics. Whereas the Americans had spurned neutralism, the Soviets now welcomed it as an ally in reducing Western influence in the Middle East. Soviet propaganda championed the Bandung Conference, billowed monotonous warnings against American aggression, and embraced the Arab nationalism it had only recently denounced. Trade accords with Egypt, Syria, and Yemen followed, as did promises and driblets of economic aid—all part of a Soviet "economic offensive" in the Third World, cried Washington. The 1955 arms deal with Egypt actually constituted the first significant Soviet activity in the Middle East. In early 1956 the Soviet Union agreed to help the Egyptians initiate atomic research and began to train some Egyptian military officers in Poland. Under these agreements the Egyptians did not become Soviet clients, for they were using the Soviets for their own purposes while conceding them little influence and rejecting *quid pro quos,* such as easing repression of the Communist Party. American officials knew that Egypt was not "going Communist," but they assumed that the Egyptians were too weak and too crazed by anti-Western hatreds to resist Soviet infiltration. In reading contemporary, secret American documents on the Soviets in the Middle East, the historian is struck by the numerous, consistently vague, unsubstantiated generalizations. Few specifics, few examples—only sweeping assumptions that the Soviets *had* to be plotting in some devious way to undermine Middle Eastern governments and Western interests, if for no other reason, it was thought, than that the Soviets always attempted to exploit opportunities when the West showed weakness. Through their own misguided policies Americans were actually helping to make happen what they thought must already be happening: an expansion of Soviet influence in the region.

Through early 1956 Eisenhower officials began to explore ways to weaken and isolate the state they believed had become Russia's servant in the Middle East—Nasser's Egypt. Expressions like "Soviet-Egyptian ambitions" appeared more frequently in official correspondence.[23] The American military urged Dulles to seek American membership in the Baghdad Pact. In the State Department, a new Middle East Policy Planning Group—"Omega"—began to discuss methods for rolling back Nasserite influence through covert and other operations. In late March Dulles summarized for the President how the United States would deal with Nasser. The Egyptian nationalist must be made to "realize that he cannot cooperate as he is doing with the Soviet Union and at the same time enjoy most-favored-nation treatment from the United States." American pressure would be applied through economic coercion: denying export licenses for arms shipments to Egypt, delaying negotiations for financing the Aswan Dam, foot-dragging in answering Egyptian requests for grain and oil, and postponing a decision on the CARE program for Egypt. The United States, reported Dulles, intended as well to strengthen the Jordanian government against riots and demonstrations aimed at thwarting Jordan's participation in the Baghdad Pact, increase aid to

Baghdad Pact members, and intensify efforts to mold Libya, Lebanon, Ethiopia, and Saudi Arabia into American-leaning states. Should these measures fail, "more drastic action" might be necessary. Although State Department censors heavily "sanitized" Dulles' memorandum to the President before releasing it to historians in 1979, it can be assumed that the last reference was to covert actions aimed at removing Nasser.[24]

From the Egyptian point of view, the delay in launching the Aswan Dam project became the most irksome. The Aswan or High Dam became the keystone in Egypt's development plans. Designed to store the waters of the Nile River for irrigation of desert lands, to control devastating floods, and to produce electric power, the massive dam became Nasser's dream for meeting Egypt's food needs. Since 1953, World Bank, British, and American officials had expressed interest in helping to finance the $1.3 billion project. After all, Middle Eastern peace and American goals would be served by having Egypt divert its spending from arms to economic development and shift its attention from external enemies, Israel and Britain, to internal improvements. Negotiations moved slowly because the Egyptians found some conditions interventionist. Even though the World Bank was putting up only $200 million of the total cost, Bank officials insisted that Egypt could not make other loan deals during the pay-back period and that the Bank had the right to advise Cairo on internal economic policy. The United States, which combined with Britain to offer another $200 million, sought the right to withhold portions of the grant when it deemed circumstances exceptional—in other words, if Egypt did not behave, the United States, when it wished, could simply pull out.

Although by early 1956, after the World Bank and Egypt had agreed upon some compromise language and the Eisenhower Administration was prepared to ask Congress for appropriations, the question of dam funding stalled. Dulles wanted to see some movement toward Arab-Israeli peace talks before committing American money. Interest groups rallied against Aswan Dam financing. Southern cotton growers wailed against helping Egypt produce more of its high quality cotton to compete with the American product. Jewish lobbyists demanded delay until

Egypt made peace with Israel. In Congress, inveterate anti-Communists like Senator William Knowland of California steadfastly refused to aid a nation they considered a Soviet stooge. Knowland's cause was boosted in May by Egypt's untimely recognition of the People's Republic of China. Although some analysts read the opening of Sino-Egyptian relations as Nasser's way of insuring arms imports in case the great powers, including Russia, agreed to embargo weapons to the Middle East, others thought it but another sign of Nasser's promotion of Soviet interests. But, as we have seen, even before this Egyptian decision and the rise of political opposition, Dulles had shifted toward a stronger anti-Nasser line. By mid-1956 the question was not whether to help build the dam but how to tell the Egyptians that the deal was defunct. "Egypt," American officials reassured themselves, "had disqualified itself."[25] As the President told Dulles in mid-July, "weakening Nasser" was a topic they had discussed for some time already.[26] American ambassador to Egypt Henry Byroade dissented from the prevailing sentiment in Washington. He bluntly informed Dulles that if Americans considered neutralists like Nasser "as being either in the enemy camp or as 'fellow travellers,' I fear that before too long we will begin to appear in the eyes of these people as being the unreasonable member of the East-West struggle."[27] Egypt, he warned, will turn to Russia for economic help.

On July 19, in a meeting with the Egyptian ambassador, Dulles revealed the American decision. Hussein, who like Nasser much preferred an American-backed project but who nonetheless had casually mentioned that the Soviets would fill any gap left by the Americans, reacted angrily—not so much against the loss of the money but against the way Washington treated the Egyptians. First, they had waited years. Second, they had made numerous compromises to satisfy Western requirements. Third, they were told "no" abruptly. Fourth, Dulles had condescendingly lectured the ambassador that the Egyptians, having so recently gained independence, ought to be careful about agreements with Moscow. And, last, they felt insulted by the State Department's public explanation for the withdrawal: Economically fragile Egypt was not capable of undertaking and sustaining such a huge project; in short, its credit was no good. This despite the fact that the World Bank—which, by the way, had been given no advance notice of Dulles' decision—had declared the dam a sound investment. Dulles had apparently decided to play rough, somehow thinking that such hard-knuckled economic diplomacy would move Nasser closer to the Western camp. And the blunt, direct approach would serve warning on other neutrals that they could not expect Washington to help them if they played both sides of the Cold War. Dulles even seemed to look forward to the propaganda advantage accruing to the United States if the Soviets offered aid to the Egyptians. "You don't get bread because you are being squeezed to build a dam," American officials could tell the people of Eastern Europe.[28] Nasser was in India when the bad news was delivered. He showed Nehru the telegram, "Those people, how arrogant they are," remarked the Indian leader.[29] "May you choke to death on your fury," shouted a defiant Nasser against the United States.[30]

Egypt's "flirtations with the Soviet Union" and "bluff" about a Soviet aid offer, Dulles informed the President, lay at the root of the American decision to withdraw from the Aswan project.[31] If Dulles ever thought that the Aswan decision would persuade the Egyptians to renounce their past "sins," he learned quickly how terribly wrong he was. On July 26, Nasser startled everybody but his closest aides by announcing that Egypt was nationalizing the Suez Canal Company. He would use its revenues—then about $100 million a year—to finance the High Dam his people needed for their economic development. The company, a symbol of the colonial past, was owned by the British government (44 percent) and French citizens (50 percent). Nasser also took up the Soviet offer. The Soviets eventually gave over $500 million to finance the dam, which was dedicated in 1971, one year after Nasser's death.

Dulles, who had made the decision to withdraw the Aswan offer hurriedly and without staff consultation in the State Department, had miscalculated: He had not anticipated Nasser's bold seizure of the canal, and he had thought the Soviets could not afford the Aswan project. Now Nasser was even more hostile and even more the hero to Arabs. Now the Egyptians had their hands on the oil lifeline to NATO nations. Now Soviet-Egyptian ties were strengthened. To make the prospects for stability even dimmer, it seemed likely that the Israelis would attack Egypt before the latter could import and put into place Soviet weapons from the 1955 deal. From London and Paris Eisenhower and Dulles heard that those governments were prepared to use force against Egypt. Even in America's own sphere of influence the Suez crisis apparently threatened: Panamanians, Dulles told the President, were "conniving with Nasser and apparently deciding on a policy of making the Panamanian nationalization or internationalization of the [Panama] Canal a major issue of Panama politics."[32]

In a meeting with legislative leaders on August 12, Eisenhower and Dulles explained frankly what had gone wrong and what they proposed to do. Although agreeing with Britain and France that they would be reduced to "dependence" if they lost their Middle Eastern oil and that Nasser's "stranglehold" was thus intolerable, Dulles counseled the Europeans to renounce force. Military action would only give the Soviets "unbounded opportunities to exploit." Nasser, the conferees-turned-psychologists seemed to agree, was a "dangerous fanatic" with a "Hitlerite personality." Oil was the central issue, Eisenhower insisted, and "don't you think we intend to stand impotent and let this one man get away with it."[33] What would the United State do? Dulles said he would attend the London conference of canal users to arrange international control of the Suez Canal. The conference opened on August 16. Egypt did not attend, explaining that the Suez Canal was part of Egypt and that it had never been consulted about the convening of the conference. An "international" solution to the crisis appeared doomed at the start. Indeed, through September and into October Dulles seemed to be doing no more than delaying the inevitable British and French—and Israeli—resort to military means. Not wanting to alienate American allies in Europe, Washington could think of no way to moderate Egyptian policy. In 1952, Eisenhower had galvanized public opinion by saying, "I shall go to Korea." Here in 1956, again in a presidential race, he never said, "I shall call on Nasser." At a time when revolts in Poland and Hungary exposed heavy-handed Soviet imperialism in Eastern Europe,

Britain and France by threatening Egypt were negating the propaganda value of highlighting such Soviet repression. American policy seemed paralyzed. "We are going to have a donnybrook in this area," Eisenhower told the National Security Council on October 26.[34] Indeed, three days later Israel invaded the Sinai. Although the Israelis hatched the plot with the British and French, not one of America's allies informed Washington that it had decided upon a military assault.

Eisenhower heard about the Israeli attack while campaigning, and he ranked it as an error equal to the 1949 loss of China. "All right, Foster," the President instructed his Secretary of State, "you tell 'em, goddam it, we're going to apply sanctions, we're going to the United Nations, we're going to do everything that there is so we can stop this thing."[35] Meanwhile, the British and French issued an ultimatum to Egypt and prepared to attack. Eisenhower engaged in a "trans-Atlantic essay contest" with British Prime Minister Anthony Eden while the two Western conspirators vetoed a United Nations resolution introduced by the United States that condemned Israel for launching the Suez war.[36] On October 31, British and French planes began round-the-clock bombing raids against Egypt, which retaliated by blocking the canal with sunken, cement-laden ships, and on November 5, British and French paratroopers dropped on Egyptian soil. Syria blew up oil pipelines.

Eisenhower felt betrayed by his allies. He vowed not to meet their oil needs and not to support the collapsing British pound with a loan. He would not rescue them from their folly. They could "boil in their own oil, so to speak."[37] At the same time the President did not want to wreck an important alliance. "Those British—they're still my right arm!" he remarked to an aide.[38] On the third of November, John Foster Dulles fell ill; surgery revealed abdominal cancer. The crisis would be Eisenhower's to handle now. But he could not stop the British and French invasion of Egypt on November 5—an attack that drew the world's attention away from Soviet tanks in the streets of Budapest. Eisenhower won the presidential election easily on November 6—the same day that London and Paris accepted a cease-fire. Israel agreed on November 8. American, neutralist, Soviet, and United Nations pressure, combined with political protest in Britain, had convinced the attackers to halt. Actually American actions were never so strong as American words. Dulles, for example, admitted that the proposed economic sanctions against Israel, which never went into effect, constituted no more than a "mild slap on the wrist."[39] The war ended because the British and French realized they could not gain their objective of controlling the Suez Canal and because the Israelis had in essence reached their goal of securing their border and destroying many of the Soviet weapons recently acquired by Egypt.

Through the days of fighting, American officials had worried that the Soviets would either gain stature in the Third World, because they had stood as an opponent of Western aggression, or escalate the conflict. If the United States stood fast with Britain and France, Dulles mused, "we will be looked upon as forever tied to British and French colonialist policies."[40] Eisenhower agreed that the Soviets might seize the "mantle of world leadership," despite its roughing up of Eastern Europe.[41] And would the Soviets risk igniting World War III? Early in the crisis, President Shukri Kuwalty of Syria was in Moscow imploring the Soviets to help Cairo. "But what can we do?" asked Nikita Khrushchev as he glanced at a map.[42] Geography dictated caution; so did American superiority in the region and the troubles in Hungary. The Kremlin informed Nasser that the Soviets could not assist militarily; instead they would try to rally world opinion. Only after it appeared that the attackers would call off their venture did the Soviets threaten action. In blustering letters to Washington and the belligerents on November 5, the Soviet hierarchy hinted at the use of force. Eisenhower turned down Moscow's offer to combine American and Russian troops in Egypt in a peacekeeping mission. The CIA believed that the Soviets would do little more than keep the "pot boiling" with rhetoric, because Moscow did not want to invite a general war, although it might try to stimulate a coup in Syria, or "frighten the Shah of Iran," or "upset the Nuri regime in Iraq."[43] The Soviets threatened to send "volunteers" to fight in Egypt should the invaders not withdraw, but Nasser apparently assured Washington that he would not accept such Soviet assistance.

The Egyptians probably never expected Soviet help during the crisis. Nasser knew that the Soviets would behave like any other great power in serving their own national interest if it conflicted with that of a small state. As Anwar el-Sadat later noted with disdain, Moscow had only shown an "exercise in muscle-flexing and an attempt to appear as though the Soviet Union had saved the situation. This was not, of course, the case. It was Eisenhower who did so."[44] Although the Soviets crowed that they had forced the belligerents to back down, Nasser knew better, and Egyptian-Soviet relations cooled after the Suez crisis. Nasser thanked the United States, and he talked about improving Cairo's relationship with Washington. The crushing of Hungary could not have been too far from his thinking either. The moment seemed fertile for better Egyptian-American relations, for steps toward accommodation. Yet, the United States passed up the opportunity. It refused to meet Egypt's postwar call for food, fuel, and medicine. And Washington set about to isolate Egypt even more—this time through the Eisenhower Doctrine.

After Suez, American leaders groped for a new Mideast policy. In frequent high-level meetings they concluded that United States prestige had increased in the region but American goals had suffered setbacks: Britain, France, and Israel had become disgruntled friends; NATO was weakened; Nasser's stature in the Middle East had risen because he had been victimized by Western imperialism and had fought back; Arab-Israeli tensions had deepened; the precariousness of vital oil supplies to Europe had been demonstrated; the Soviet Union's "foothold" in the region appeared firmer;[45] and, given the decline of Western power, the Middle East now resembled a "spongy vacuum" that the Soviets might fill.[46] Like 1947 and the British withdrawal from Greece, Dulles remarked simply in reference to the Truman Doctrine, the United States had to move into the power vacuum. To Eisenhower, because "the Bear is still the central enemy," the United States had no choice but to assume the role of policeman in the Middle East.[47]

But the United States needed allies for this task. The administration was taken aback by outbursts of strident anti-Americanism in Britain and France. The British, including some of the President's World War II friends, resented Washington's unwillingness to provide a loan to shore up the sinking pound. In France, service station attendants refused to pump gas into American cars, and taxis would not pick up Americans. "Our country is on the way to drawing down on itself, in Europe, a bitterness of feeling comparable to the bitterness felt in the 1920s," wrote an observant American scholar from Geneva.[48] Britain and France, reported CIA Director Allen Dulles, "were in a highly psychopathic state which promised to become worse with the onset of cold weather."[49] Eisenhower feared that the Western alliance was breaking apart; its repair became top priority. His anger with the British soon subsided; day after day he tempered his criticism of London. "We must face the question, what *must* we do in Europe and then the question, how do we square this with the Arabs."[50] Thus the Eisenhower Administration approached its Mideast policy from the perspective of its European policy. Or, put another way, looking through a European or Cold War lens, Eisenhower officials continued to shape a policy for the Middle East that left little room for accommodation with Arab nationalism or with its ambitious spokesman, Gamel Abdel Nasser. As the President said in late November, "we should give the British every chance to work back into a position of influence and respect in the Middle East."[51] Few Arabs, Nasserite or otherwise, welcomed the restoration of British influence.

Why not join the British in the Baghdad Pact? The Joint Chiefs of Staff urged American membership, but the State Department urged caution. Many Arabs considered the pact a device to extend British influence. Even Saudi Arabia was highly critical of the alliance, and because King Saud was evolving as America's "counterpoise" to Nasser, membership would jeopardize important relations with an emerging ally.[52] As well, Iraq waxed angrily anti-Israel: The American government could hardly align with such an outspoken foe of the Jewish state. It seemed likely, too, that Congress would permit membership in the pact only if the United States formed a similar military alliance with Israel— a step sure to alienate most Arabs even more.

With little settled, yet worn low by the demands of the fall crisis, the President departed Washington in late November for relaxation and golf at his favorite spot in Augusta, Georgia. Before Eisenhower's return on December 12, a healthier John Foster Dulles had become active again, and under his guidance a new policy began to emerge. Senior officials recommended that Nasser's Egypt be made to "behave"; they did not like its "playing both sides to the middle"; and, when American pressure was applied, he had to be stopped from turning toward the Soviet Union.[53] "We regard Nasser as an evil influence," the President concluded.[54] Having ruled out overtures to or negotiations with Egypt, the Secretary of State presented via telephone three alternatives to the President. First, American membership in the Baghdad Pact. Second, organization of a new group of allies. Third, a nation-by-nation approach utilizing foreign aid (and armed forces if necessary) authorized by congressional resolution. Eisenhower, who said he liked to "carry two strings in the bow," liked numbers one and three. "If we should get Saudi Arabia and Lebanon to adhere to the Pact we could go in with them and that would be wonderful." Dulles thought not: The Saudis would not join; and American Jews would lobby Congress against American participation. But the pact was a "defense against Communism," insisted the President. Not quite so, replied Dulles, for it had also become an "instrument of Arab politics."[55] The President then gave up on his flirtation with membership in Baghdad. He told Dulles to compose a congressional resolution to exercise option three.

By January a resolution and a speech were ready—hammered out by State Department writers through several drafts, with close questioning and editing by the President. On the afternoon of New Year's Day 1957, Eisenhower convened a large bipartisan congressional leadership meeting in the White House. He and Dulles, uttering the word "vacuum" repeatedly, suggested that the Soviets might try to recover from their "deterioration" in Eastern Europe by seeking "victory" in the Middle East. The world had to be put "on notice" that the United States intended to deter Soviet "aggression" in the vulnerable region, for the "loss of the area would be disastrous to Europe because of its oil requirements." Saudi Arabia, Iraq, Iran, Libya, and Lebanon had to be reassured about their security the way Greece and Turkey were buttressed by the Truman Doctrine a decade earlier. Throughout the nearly four-hour meeting, the President and Secretary were extremely vague about the nature of the Soviet threat. At times they spoke as if Soviet troops would pour into the Middle East. Except for mention of Soviet military shipments to Egypt and Syria, no American official provided a detailed, substantiated account of Soviet intentions, activities, or capabilities. And nobody seemed to want to ask. The assumption prevailed that the Soviets would exploit any chaotic area. Dark images of "international Communism" seemed adequate, and exaggerations became legion.[56]

On January 5, the President spoke to a joint session of Congress and repeated the anti-Communist theme that had become a central feature of Cold War politics. "Aggression," both direct and indirect, and "international Communism" spearheaded by the Soviet Union were two of the address's most frequent utterances. Eisenhower requested a congressional resolution granting him authority to employ American armed forces in the Middle East "as he deems necessary" to help nations which request aid "against overt armed aggression from any nation controlled by international Communism," to undertake military assistance programs, and to provide economic assistance.[57]

Even before the Eisenhower Doctrine speech and the Middle East Resolution went into final draft, critics within the administration questioned the heavy anti-Soviet emphasis and predicted that many Middle East nations would read the new policy as a Western thrust into the region—this time through an American claim to the right of intervention to stop Communism. When one CIA officer was asked to prepare explanations of the new doctrine for Arab chiefs of state, he indignantly replied that "we can't afford to associate ourselves with every lunatic scheme that comes along."[58] In the Congress the resolution ran into "a great buzz saw of questions."[59] Did not the president already possess the authority he was requesting? And, if he did, did not the request confuse the issue of presidential power? And, if he did not possess the authority—especially to use armed force—was not Congress, in agreeing to the resolution, abdicating its power to declare war, granting "a predated declaration of war?"[60] Eisenhower had little interest in these legal questions; constitutional musings should not impede the necessity of containment. Other questions reached the White House: Would not the introduction of American arms into the Middle East only aggravate local problems? asked critics like Senator J. William Fulbright. Would not Arabs be tempted to use the arms against Israel, should another war break out? Yes, these were worries, the administration answered, but the United States must be willing to take such risks on behalf of anti-Communism.

Was there a Soviet/Communist threat to the Middle East? The administration's answers led to confusion. Dulles and others admitted that there really was no direct Soviet military threat; there were no Soviet troops in the Middle East, although some faced Iran and Turkey. Would the Soviets actually precipitate a showdown with the United States by attacking those two nations, both American friends? Not likely, replied Dulles, but remember Korea in 1950, and, anyway, the always aggressive Soviets had to be told now what the risks were if they ever moved against the Middle East. Were there any countries in the area "controlled by Communism?" No. Was Egypt, the nation which had banned the Communist Party, a Communist state? No, but it was taking Soviet arms, including MIG fighters. Was Syria Communist dominated? No, but it received assistance from the Soviet Union. And Soviet subversion was a danger throughout the Middle East. Then why did not the resolution mention subversion? Because Arabs would read it as an American exaggeration designed to facilitate United States intervention. Senator Fulbright grew frustrated. The White House, he complained, "asks for a blank grant of power over our funds and armed forces, to be used in a blank way, for a blank length of time, under blank conditions with respect to blank nations in a blank area. . . . Who will fill in all these blanks?"[61]

To Eisenhower and Dulles, the senator from Arkansas was unnecessarily complicating the matter. Although admitting that the Communist threat was difficult to define or substantiate, they insisted that it was important to declare a policy of deterrence for the Middle East, to implant a "keep out" sign, to initiate a police patrol, to issue a warning for the future. In the end, the Eisenhower Doctrine was more propaganda than anything else; it bought time while the United States built influence in the area. It was confused and vague enough to meet any number of contingencies, including a Nasserite threat. Yet it did not address the major obstacles to peace in the Middle East: Arab-Israeli tensions, Palestinian refugees, and profound Arab anti-Western sentiment. American officials were not ignorant of these problems, but they saw them as long-term; the immediate need required emplacement of a protective shield over the vulnerable Middle East to deflect an inexorable Soviet threat. Although anti-Communist rhetoric dominated, it was no secret that Egypt, seen as Moscow's tool, was also a target. The Congress agreed; the House passed the resolution—after some tinkering with language—on January 30, by a 355 to 61 count; the Senate followed on March 5, with a 72 to 19 vote.

"With all the subtlety of temperance crusaders in a distillery," Washington set out to woo Middle Eastern states to its anti-Nasser, anti-Soviet enterprise.[62] But many Arabs saw neither a direct nor an indirect threat from Soviet/Communist aggression; they saw Western meddling and sphere-of-influence building; they sought an American declaration against Israeli, British, and French aggression; and they denied American depictions of a "vacuum," for to them, Arab nationalism had already filled it. Syria, Egypt, and Jordan rejected the Eisenhower Doctrine. The Jordanian premier said Jordan did not wish to exchange one master (Britain) for another (the United States). Saudi Arabia and Iraq were lukewarm. Non-Arab Iran applauded the new policy. The only Arab state to endorse the doctrine fully was the half-Christian, half-Muslim state of Lebanon. Israel mildly accepted the doctrine, but it advised that the threat in the Middle East was rabid Arab nationalism, not Soviet aggression; Israel did not look kindly upon the prospect of America's strengthening anti-Israeli forces in the region. Israel also made the Eisenhower Doctrine's going more difficult by refusing to withdraw its troops from the Gaza Strip; not until early March, after the United States threatened economic sanctions, did the Jewish government pledge withdrawal. Two characteristics were shared by these critics of the new Middle Eastern policy: None had been consulted by the United States before Eisenhower made his speech; and all thought American fears of Communist inroads were unjustified. Perhaps the greatest beneficiary of the Eisenhower Doctrine was the Soviet Union, for it was able to associate with Arab hostility to the American policy.

Dulles and Eisenhower fixed on Saudi Arabia as one Arab state that might help make the new doctrine work; they had long believed that the Saudis could be cultivated and drawn closer to the West, for King Ibn Saud worried with them about Nasser's radicalism. Perhaps Saud could be induced to break with Cairo; he was certainly "one of the most important anti-communist assets in the area," Dulles remarked.[63] Eisenhower invited the Saudi monarch to visit the United States in late January and early February. He arrived with an entourage of wives and sword-bearing bodyguards. In Washington, American leaders explained the dangers of international Communism, while Saud talked about the continued British presence and Israeli behavior. But a deal was struck: Saudi Arabia expressed support for the Eisenhower Doctrine and pledged a new five-year American lease of the airbase at Dhahran; in return, Eisenhower promised aid to the Saudi armed forces. In a meeting between the King and the President, with only an interpreter present, Saud urged Eisenhower to invite both Nasser and the King of Syria to Washington. "He said that he believed great good could come of such visits," Eisenhower noted in a memorandum. "I had not expected this one and so I stalled a little bit. . . . He hastened to interject that he was certain these people did not lean so much toward the Soviets as we had thought and they would like to re-establish their ties with the West." But the realistic King recognized the difficulty of an overture to Nasser, for he "knew something about our political situation, at least that we had a lot of Jewish voters in this country."[64]

If Eisenhower had taken the political risk in 1957 and invited Nasser to begin talks (they would meet once only briefly in 1960), future troubles for American foreign policy in the Middle East might have been tempered, and the sensible American abandonment of the Eisenhower Doctrine might have come sooner. Making the doctrine work became nearly impossible. The administration sent military and economic aid and emissaries to the region to build an anti-Communist front, but it could never organize the anti-Soviet Arab bloc it sought to build.

The first test came in the Jordanian crisis of April 1957. Anti-United States, pro-Nasser politicians, stimulated to action in part by the announcement of the Eisenhower Doctrine, won elections and put pressure on King Hussein to loosen Jordan's ties with Britain and the United States. Hussein cried that international Communism threatened his monarchy; and he appealed to the United States to invoke the Eisenhower Doctrine. The American Sixth Fleet moved into the Eastern Mediterranean; Washington pledged millions of dollars in aid. Despite the fact that Jordan was threatened neither by Soviet aggression nor local Communists, Hussein emerged triumphant and drew closer to the United States. In the first application of the Eisenhower Doctrine, then, the target was Nasserite, Arab nationalism, not Communism.

Next came a contest with Syria, where leftist, pro-Nasser factions gained power and negotiated greater Soviet assistance. When an abortive CIA plot to put a pro-Western faction in power was discovered by the Syrians, relations between Damascus and Washington plummeted. Syrian politics became rife with turmoil and intrigue. The Eisenhower Administration faced a predicament: The doctrine said that the United States would intervene to prevent a Communist takeover when invited by the threatened nation. Obviously no invitation would be forthcoming in this case. Dulles advised the President in late August 1957 not to declare Syria yet a victim of "international Communism" under the doctrine, because "we cannot yet make a clear judgment as to the actual extent of Communist penetration." Nonetheless, Dulles went on, a familiar scenario was unfolding, suggesting Syria was becoming a Soviet satellite:

> There is evidence in Syria of the development of a dangerous and classic pattern. The Soviets first promise and extend aid, military and/or economic. With this aid they promote the control of any positions by pro-Soviet persons. The end result sought is that the country will fall under the control of International Communism and become a Soviet satellite, whose destinies are directed from Moscow.[65]

From Saudi Arabia's King Saud came a quite different message: Ambitious Army officers, not Communists, lay behind the pro-Nasser thrust of Syrian politics. When Eisenhower relayed Saud's message to congressional leaders in a White House meeting, he remarked that General Charles de Gaulle had once told him that no true Frenchman could be a Communist. Ike distrusted such generalizations. Interpreting the Syrian crisis as a possible Communist bid for power and hence a threat to the West's oil supply, he defined American policy: encouragement to neighboring Turkey to put pressure on Syria and increased shipments of American arms to Jordan, Lebanon, Iraq, and Turkey. By the fall, a hot war between Turkey and Syria seemed likely. That danger persuaded many Arab states to rally behind the Arab nationalists in Damascus. Egypt sent troops to Syria. Even Saudi Arabia pledged support to the Syrians if the non-Arab, American ally Turkey attacked. When in October, Lebanon, Iraq, and Jordan joined Saudi Arabia in backing Syria, it became evident that Washington's policy of inducing Turkey to weaken the pro-Nasser government in Syria had actually stimulated a higher degree of Arab solidarity against the United States. In February 1958 Syria and Egypt merged as the United Arab Republic (UAR). Nasser apparently waxed unenthusiastic about the arrangement, but accepted Syria's initiative for fear that rejection would weaken Arab nationalism. He quickly banished the Communist Party of Syria. As it turned out, the Syrian leaders the United States had tried to dislodge were actually anti-Communist.

The Eisenhower Doctrine, in this second test, had ended up, once again, at odds with Arab nationalists who had no intention of permitting Communist gains, who were as determined as Americans to deter Soviet domination of the Middle East, and who were largely reaching for Soviet military and economic assistance to counter the West and Israel. In the Syrian crisis, the British analyst Patrick Seale has written:

It was a situation in which the United States could be said to have been mesmerized by a monster of its own creation. The danger of a Soviet take-over had been so explicitly heralded, a battle-drill of such precision had been prepared, resources of such magnitude had been deployed to guard against a surprise attack that, now that the enemy appeared to have struck, action could no longer be avoided.[66]

As with Egypt, so with Syria—the United States made no attempt to deal directly with Syria, to explore ways of tempering anti-Western views, or to discover just how safe Syria was from Communism.

Lebanon provided another test for the Eisenhower Doctrine. In mid-May 1958, President Camille Chamoun, a Christian in the multi-confessional nation, precipitated a domestic crisis by maneuvering for a second term, despite a constitutional prohibition against it. The ambitious Chamoun had already used CIA money in elections to undermine Muslim candidates. His many critics believed that he was threatening the National Pact—the agreement that maintained some political stability in the fragile country by allocating offices to the various religious groups. When civil war erupted, it was marked by anti-Americanism, for Chamoun had been an early and enthusiastic champion of the Eisenhower Doctrine. In Tripoli a mob burned the United States Information Agency library. Chamoun appealed to Washington for help, citing United Arab Republic assistance to his opponents and Radio Cairo's relentless propaganda. Nasserism and Communism, he cried, were attacking Lebanon. But, as American Ambassador Robert McClintock reported from Beirut, the "authentic indigenous element of opposition" to Chamoun "cannot be dismissed" as Nasserite.[67] On May 13, a panicky Chamoun asked the British, French, and Americans to "consider" landing their armed forces to quell the rebellion and preserve the nation's independence.[68]

In Washington that evening, Eisenhower huddled with his key advisers. They agreed that the United States should assist a Middle East friend, and they recognized that the Eisenhower Doctrine did not directly apply, because the case that the UAR was controlled by international Communism or that the UAR had attacked Lebanon would be difficult to make. Still, under the general language of the congressional resolution that the preservation of the independence and integrity of Middle Eastern nations was vital to world peace, the United States could justify intervention. But more, advised Secretary Dulles, "we would have to accept heavy losses, not only in Lebanon but elsewhere" if the United States stood aside. For Dulles, and apparently for others at the White House meeting, the Lebanese crisis was but one example of "Communist methods of warfare throughout the world." In this remarkable misreading of the turmoil in Lebanon and of McClintock's diplomatic reporting, Dulles mentioned similar contemporaneous Communist "tactics" in Venezuela, Burma, and Indonesia. That very day demonstrators had attacked Vice President Richard Nixon's automobile in Caracas.[69] Within this broad context of a Communist threat, Eisenhower decided to honor Chamoun's request for armed forces when and if the Lebanese leader made it. The American military was put on alert.

As the crisis in Lebanon simmered down and Nasser urged compromise, American leaders felt trapped. They seemed powerless to manage Lebanon's faction-riddled politics, yet they had given Chamoun a certified check that he could cash at any time. Intervention, American officials agreed, would solve little; indeed, it would "represent a victory for Nasser" because of intensified anti-Western feelings that would follow and inflame the Middle East.[70] "How can you save a country from its own leaders?" wondered Eisenhower. "There seemed nobody on whom we could pin our hopes."[71] Still, as Dulles explained, America had to act if asked; otherwise Lebanon would become "a dependency of the UAR and perhaps, realistically, of the USSR."[72]

On July 14, ugly events in Iraq forced the question. Pro-Nasser army units swiftly seized power in Baghdad and assassinated the royal family. Fearing a Nasserite "brush fire" sweeping across the Middle East, engulfing the American friends of Jordan, Saudi Arabia, and Lebanon, Eisenhower accepted Chamoun's urgent request for troops.[73] Although they knew the evidence was very thin, Eisenhower and Secretary Dulles mentioned again and again, in high level meetings, that the United Arab Republic and the Soviet Union were fomenting the new troubles. Nasser is "so small a figure, and of so little power, that he is a puppet," remarked the President, "even though he probably doesn't think so."[74] Nasser was seeking to control petroleum supplies so that he could "destroy the Western world."[75] Nasser and Khrushchev, then, shared a common purpose, and the "Reds" were "careful not to show [their] hand" in using "Nasser to play the game."[76]

On July 15, to the surprise of bikini-clad sunbathers and village workers, thousands of Marines rushed a Lebanese beach near Beirut. "What you come here for? To start war?" asked an unfriendly Lebanese boy.[77] But, on the whole, the reception was peaceful. One Marine gazed over the tranquil scene: "It's better than Korea, but what the hell is it?"[78] American troops first secured the airport and then took up defensive positions in Beirut. During the occupation, they received some fire from rebels but suffered only two deaths from accidental shootings by other Marines. American armed forces were withdrawn from Lebanon in late October after a new government, arranged by Deputy Under Secretary of State Robert D. Murphy, took power—without Chamoun.

The Eisenhower Administration claimed that it had saved Lebanon, but critics decried a renewal of gunboat diplomacy. If the Soviets were involved as charged, then go to a summit meeting with Khrushchev to set limits on their behavior; if Nasser was engineering the revolts, then talk with him, or impose economic sanctions on Egypt and Syria; if Radio Cairo was stirring up turmoil as charged, then jam it. But not a military intervention, advised dissenters. If the invasion was designed to show Nasser that he could not depend upon the Soviet Union in a crisis—in fact, the Soviets did little more than crank up their propaganda machine—such thinking ignored history, critics asserted, for Nasser had already learned in the 1956 Suez crisis that Moscow would not take risks. Doubters like Senators Fulbright and Hubert H. Humphrey pointed to civil wars and Arab nationalism as the sources of regional instability and explained the inappropriateness of military methods. During the American Civil War, former Secretary of State Dean Acheson remarked, the "Confederates looked to England for help and the *Alabama* was built in British shipyards."[79] Did that mean the Confederates were dominated by England? Revolutionaries always seek help from outsiders; Arab rebels sought it from Nasser and Moscow but got little. Opponents of the Lebanese venture demanded evidence of UAR or Soviet instigation or control of Arab insurgents, but got none. Nor did the United Nations Observation Group report any. And the emissary Murphy concluded: "Much of the conflict concerned personalities and rivalries of a domestic nature, with no relation to international issue. Communism was playing no direct or substantial part in the insurrection. . . ."[80] The Iraqi and Lebanese governments, it seems clear, were threatened not by Nasserism or Communism from afar but by some of their own people at home.

After the Lebanese intervention, the Eisenhower Doctrine would seldom be heard from again, although exaggerations of the Soviet/Communist threat remained strong. Such habits of thought proved tenacious even in the face of Middle Eastern realities that generated problems for the United States: bitter intra-Arab politics, Arab-Israeli conflict, anti-Western sentiment, neutralism, homeless Palestinians, and Nasserism and pan-Arabism. The Eisenhower Doctrine contested rather than engaged Arab nationalism; it failed in its objective of isolating Nasser; and it did not thwart "international Communism," because such a thing had never visited the Middle East in the first place. By the end of the Lebanese affair, the United States, not the Soviet Union, starred as the villain in the Middle Eastern drama. The reasons for this tragedy lie not only in the hostilities generated by Arab-Israeli disputes but in many Eisenhower-Dulles misjudgments: Nasser could be scolded like a child and contained; Arab nationalism was serving sinister Soviet goals; local conditions were insufficient to impede Communism; peace could be had without pressing Israel to be cautious; the Baghdad Pact would work; military methods—alliances, arms, and interventions—were suitable; and the Eisenhower Doctrine would protect American interests and win friends. Some have argued that the region's problems had become so daunting and intractable that any American policy under any President would have failed. But such a conclusion obscures American policymakers' repeated misreadings and mishaps that made a bad setting worse, destroyed the possiblity of creating an improved environment in which small steps could be taken, and intensified the danger to American lives and interests. Perhaps, if American leaders had dealt with Middle Eastern realities rather than with Communist ghosts, they could have come, at a minimum, to understand the region's profound stirrings as indigenous rather than imported, and might have devised a patient, sensitive, and tolerant policy. But they did not, because they had become Cold Warriors, hyperbolic in their depiction of the Communist threat.

# 10

## Bearing the Burden: John F. Kennedy and the Communist Menace

One question quickly jumps out in front in any discussion of the foreign policy of President John F. Kennedy in the early 1960s: Why were there so many serious, potentially catastrophic contests and crisis? The foiled Bay of Pigs invasion, deadlock at the Vienna summit, Berlin crisis, Laotian crisis, escalation in Vietnam, blunt bickering with France's Charles de Gaulle, Congo crisis, missile race, space race, and the Cuban missile crisis—all cascaded through world affairs in just 1000 days. How do we explain this frightening set of confrontations, this competition and brinkmanship that seemed at times to court nuclear cremation?

Getting at the well-springs of the Kennedy foreign policy requires overcoming some heady obstacles. Many of the records for the Kennedy years remain security-classified and hence closed to researchers. We must do our reconstructing without all of the parts; some papers, such as those generated by the CIA, may never be released to historians. Very slowly and incompletely this difficulty is being reduced as the John F. Kennedy Library and federal agencies open documents for research under the mandatory review process and the Freedom of Information Act.

Another obstacle stems from the adulatory, best-selling memoirs and biographies penned by Kennedy's aides. They cover his blemishes with the cosmetic cream of hero worship, perpetuating a Kennedy legend. But that legend now wobbles in the face of a fuller documentary record. The books and fawning messages of Arthur M. Schlesinger, Jr., Kenneth O'Donnell, Pierre Salinger, and Theodore Sorensen have been subjected to searching study, and the Kennedy who emerges appears less and less to be the cool-headed, far-sighted crisis manager they have depicted.

We must also grapple with the question of rhetoric. JFK said so many grand things so elegantly. One can easily get caught up in the eloquent phrasing and noble appeals to human uplift and overlook contradictions between word and deed or the coercive components of American foreign policy. Kennedy also said so many hackneyed things so superficially, both publicly and privately. He often spoke of the "Communist offensive" or the "free world" as monoliths, ignoring complexities. Kennedy defenders like Schlesinger ask us to dismiss such statements as mere political rhetoric or as State Department "boilerplate." It seems more sensible to conclude that Kennedy said what he meant and meant what he said. Besides, what he said is what the Soviets and others heard.

Ambiguity also dogs us. Kennedy's foreign policy was a mixture of sincere idealism and traditional anti-Communist fervor. The President sent Peace Corps volunteers into needy and appreciative villages in Latin America to grow food. But he also sent the Green Berets into Southeast Asia, where they helped destroy village life. We are left, then, with part hawk and part dove — an administration which had serious doubts about the cliches of the Cold War but never shed them.

The assassination poses another interpretive obstacle. Kennedy admirers have asked us not to judge him by his accomplishments but rather by his intentions, for, they have argued, had he not been removed from his appointed journey so tragically in 1963, his good intentions would have reached fruition. The first 12 to 18 months of any presidency constitute a learning or trial period, and truly successful Presidents need two terms to achieve their goals. Put another way, Kennedy would have gained experience, been chastened by crisis, and become educated to follow a more temperate and less traditional Cold War diplomacy had he lived and won re-election in 1964. We cannot be sure, but we do know what he *did* in the thousand days of his administration to help stimulate a rash of diplomatic crises in such a very short time.

Finally, as historians we must contend with the difficulty of emphasis, of choosing between different roads that lead to an understanding of the Kennedy years. We can, for example, emphasize the world the President faced—a bellowing Nikita Khrushchev, independent-minded leaders like de Gaulle, neutralism, national liberation movements, guerrilla wars, and many newly freed, internally divided, and unstable states in the Third World. We can, in other words, stress that the crises of the early 1960s derived from external forces, perhaps beyond any man's comprehension or mastery.

Or, for a second road, we might highlight, as Schlesinger has done in his book, *Robert Kennedy and His Times*, the decision-making environment in Washington, D.C., that brought troubles to the President's diplomacy. Schlesinger blames the CIA as a runaway, out-of-control agency that did not tell the President what it was doing abroad to undermine and topple governments. He chides, furthermore, conservative congressmen and senators who gave Kennedy little room to relax the Cold War, and State Department bureaucrats who obstructed imaginative change. And he censures the military—the Joint Chiefs of Staff, and warriors like Curtis Lemay, Edwin A. Walker, and Maxwell Taylor—for advocating an aggressive, hawkish foreign policy. In short, suggests this point of view, Kennedy should be praised for keeping the adventuresome hawks in check, for curbing their warmongering appetite for even more crises.

A third route we might take is to look at the man and his advisers, to delineate their ideas, their styles, their goals, for, although the world was a menacing one, it had long been so, and the noise from Moscow had not really changed in kind, but in volume. And, although Kennedy probably did restrain some militarists, there was enough activism, militancy, and zeal for Cold War victory in Kennedy himself and in his close advisers to suggest that historians need to fathom them and their reading of the national interest to render a thorough accounting of 1960s foreign policy.

Kennedy once said that one man "can make a difference."[1] We may quibble with such an emphasis on individuals in history. It obscures the basic continuity in American foreign policy, the traditional expansionism and interventionism spawned by a liberal ideology and by the real economic and strategic needs of a large, industrial power with global interests. Kennedy did not represent a sharp break with the past or a uniqueness in the fundamental tenets of American foreign policy. Yet the different methods he chose to meet the Communist threat, the personal characteristics of his diplomacy, did matter in heating up the Cold War, threatening nuclear war, and implanting the United States in the Third World as never before.

What made Kennedy's foreign policy tick? First, the historical imperatives of experience and ideology which linked Kennedy's generation to a past of compelling lessons. Second, the conspicuous style, personality, and mood of the President and his advisers, who were determined to win the Cold War by bold action. And third, counter-revolutionary thought, best summarized by the phrases "nation-building" and "modernization," demanding a high degree of activism in the Third World. "The difference between the Kennedy and Eisenhower administrations," Special Assistant Walt W. Rostow has written, "is not one of 180 degrees. The difference was a shift from defensive reaction to initiative. . . . "[2]

The first explanation, the power of historical conditioning, derives from the truism that we are creatures of our pasts, that long-held assumptions, traditional behavior, and habits tug at us in the present. John F. Kennedy and his advisers were captives of an influential past. They constituted the political generation of the 1940s, and they often flashed back to that decade for reference points. Many of them came to political maturity during World War II and the early years of the Cold War. Kennedy himself served during the Second World War on PT 109 and was elected to Congress in 1946, just a few months before the enunciation of the Truman Doctrine. Kennedy and his advisers were members of what we might call the "containment generation," which enjoyed what they considered the triumphs of aid to Greece and Turkey, the Marshall Plan, the Berlin Blockade crisis, NATO, and Point Four.

They also suffered the frustration of Jiang Jieshi's collapse in China and the stalemate of the Korean War. When asked in 1963 whether he would reduce aid to South Vietnam, the President replied that he would not. "Strongly in our mind is what happened in the case of China at the end of World War II, where China was lost. . . . We don't want that."[3] Indeed, in 1949 Congressman Kennedy had blasted President Truman for the "loss" of China and had also criticized Franklin D. Roosevelt for selling out to the Russians at the Yalta Conference.

Kennedy's "containment generation" imbibed several lessons from the postwar years of the Soviet-American confrontation: toughness against Communism works; a nation must negotiate from strength; precautions must be taken to avoid compromises or sell-outs in negotiations; Communism was monolithic; Communism was a cancer feeding on poverty and economic dislocation; Communism had to be contained through counter-force on a global scale; revolutions and civil wars were usually Communist-inspired or exploited by Communists; and a powerful United States, almost alone, had the duty to protect a threatened world. Some of these lessons were exaggerated, ill-defined, superficial, or downright mistaken, yet a generation of Americans committed them to memory in the 1940s. That generation, once in positions of governmental authority in the 1960s, constantly looked back to that earlier decade for inspiration and guidelines.

Kennedy also tapped that generation for his administration's personnel. The carry-over of ideas and public servants from the Truman period to the Kennedy era is striking. Secretary of State Dean Rusk had been an Assistant Secretary of State under Truman. McGeorge Bundy, Kennedy's bright, persuasive Special Assistant for National Security Affairs, like Kennedy a little over forty years old, was Captain Bundy during World War II, had a hand in the Marshall Plan, and had developed close relations with Henry L. Stimson and Dean Acheson. Bundy revealed that "he had come to accept what he had learned from

Dean Acheson—that, in the final analysis, the United States was the locomotive at the head of mankind, and the rest of the world the caboose."[4]

Bundy's forty-four-year-old assistant, the Massachusetts Institute of Technology economist and grand theorist Walt W. Rostow, had served in the wartime Office of Strategic Services (OSS) and had worked on postwar European reconstruction as a State Department officer. His wife once observed that the Kennedy advisers were "the junior officers of the Second World War come to responsiblity."[5] Rostow himself wrote in the early 1960s that the "first charge of the Kennedy Administration in 1961—somewhat like the challenge faced by the Truman Administration in 1947—was to turn back the Communist offensive. . . . "[6] Another White House assistant, especially concerned with Latin American affairs, was Arthur M. Schlesinger, Jr., part of the Harvard contingent which trooped to Washington to serve the new President. As the forty-three-year-old historian declared on CBS's *Face the Nation* in 1960, the new administration "will have to come up with new initiatives and ideas comparable to the great creative conceptions of the 1940's."[7] For Schlesinger, the 1940s were active years as a member of the OSS and as a young Democratic party liberal who also worked for the Economic Cooperation Administration.

Other members of the "containment generation" took new posts in Washington to help Kennedy: Chester Bowles, John Kenneth Galbraith, Roger Hilsman, Robert McNamara, Adlai Stevenson, and Maxwell Taylor. Other more experienced hands like W. Averell Harriman, Clark Clifford, A. A. Berle, Charles Bohlen, and Robert Lovett came back. Paul Nitze, author of NSC-68, became Assistant Secretary of Defense for International Security Affairs. Even the ultimate in Cold Warriors, the inveterate Dean Acheson, advised Kennedy on European affairs. In 1959, when Acheson was vociferously attacking George F. Kennan for the latter's proposals for disengagement from Central Europe, Walter Lippmann complained with insight that Acheson and his rigid types were "like old soldiers trying to relive the battles in which they won their fame and glory. . . . Their preoccupation with their own past history is preventing them from dealing with the new phase of the Cold War."[8]

Reflecting on some of the foreign policy mistakes of the 1960s, Clark Clifford, White House aide to Truman and later Secretary of Defense under Lyndon Johnson, admitted that "I am a product of the Cold War. . . . I think the Truman Doctrine, the Marshall Plan, and NATO saved the free world. I think that was one of the proudest periods of our history. But I think part of our problem in the early nineteen-sixties was that we were looking at Southeast Asia with the same attitudes with which we had viewed Europe in the nineteen-forties. . . . The world had changed but our thinking had not, at least not as much as it should have."[9]

The ideas of the generation of the 1940s were molded not only by their immediate experience with the early Cold War but by their inherited assumptions from the 1930s—assumptions which blended with and explained Cold War crises. The lessons of the 1930s were widely shared: aggression cannot go unchallenged; military force had to be used decisively; economic depressions breed totalitarianism and war; there was little difference between Nazi Germany and Soviet Russia or between Hitler and Stalin (see "Red Fascism" in Chapter 1); and fanatical dictators, to maintain their power at home, become aggressors and cannot be moved by reason. John F. Kennedy submitted a senior thesis to Harvard and published it in 1940 as *Why England Slept*. Its theme was direct: The English revealed weakness before the Nazi threat and should have employed force. For Kennedy's generation, the Munich agreement became the Munich "syndrome" or lesson, a vivid example of the costs of softness. In his televised address to the nation on October 22, 1962, in the terrible throes of the Cuban missile crisis, Kennedy summoned that historical legacy for a rationale: "The 1930's taught us a clear lesson: aggressive conduct, if allowed to go unchecked and unchallenged, ultimately leads to war."[10]

The Kennedy team, then, came to office with considerable historical baggage. They felt, too, that the Eisenhower Administration, unimaginative and stale in the 1950s, had let American power and prestige slip and had thereby permitted the lessons of the 1930s and 1940s to slump from underwork and passivity. "We have allowed a soft sentimentalism to form the atmosphere we breathe," charged Kennedy. "Toughminded plans" were required.[11] They craved triumphs like those over Nazism and Stalinism. They charged, with remarkable exaggeration, that Eisenhower was unwilling to enter the new battleground of the Cold War, the Third World—that he was consigning it to Communism without a fight. "I think it's time America started moving again," proclaimed John F. Kennedy.[12] "Our job was to deal with an automobile with weak brakes on a hill," recalled Walt Rostow, with his usual exaggeration. "It was slowly sliding backward. If we applied enormous energy, the car would begin to move forward and in time, we would get it up to the top of the hill."[13]

The presidential campaign of 1960 demonstrated the ingrained nature of the Cold War's history. Richard Nixon and Kennedy differed little in foreign policy views. Nixon, too, was part of the "containment generation," also elected to Congress in 1946. Throughout the 1950s, Kennedy had proven his Cold War credentials, calling for larger military appropriations than Eisenhower wanted. In 1956 Kennedy considered Vietnam the "finger in the dike" of Communism.[14] In 1960 his Cold Warriorism surfaced as he and Nixon escalated their claims to the inherited wisdom of the past. Although Eisenhower had just suffered a set of diplomatic blows with the U-2 affair, the collapse of the Paris summit meeting, Castro's rise in Cuba, accelerated war in Indochina, an adverse balance of payments, and the forced cancelation of a trip to Japan, Kennedy had to admit when pressed on *Face the Nation* in April 1960 that he endorsed most of Eisenhower's policies, except the apparent neglect of developing nations. As David Halberstam has correctly concluded, the Kennedy people "were not dissenting from the assumptions of the Eisenhower years, but pledged to be more effective, more active, to cut a lot of the flab off."[15]

In his campaign speeches, Kennedy, who said he did not mind being called Truman with a Harvard accent, hammered away on the issue of Cuba and the Cold War. He urged pressure on Castro and aid to Cuban rebels to overthrow him. "I wasn't the vice president who presided over the communization of Cuba," he declared. "I'm not impressed with those who say they stood up to Khrushchev when Castro has defied them 90 miles away."[16] In August, he announced: "I think there is a danger that history will make a judgment that these were the days when the tide began to run out for the United States. These were the times when the Communist tide began to pour in."[17] A month later he embellished his rhetoric: "The enemy is the Communist system itself—implacable, unceasing in its drive for world domination. For this is not a struggle for the supremacy of arms alone—it is also a struggle for supremacy between two conflicting ideologies: Freedom under God versus ruthless, godless tyranny."[18] John Foster Dulles could not have said it better.

Was all of this mere hyperbolic, campaign rhetoric? It cannot be dismissed so easily. Such utterances, heard over and over again from the lips of the Kennedyites, represent the historical imprint upon a generation of Americans. History was not so much a way of learning, as a soothing way of making sense out of complex events. The historical imperative helped compel them to try to move the Cold War from stalemate to American victory. Theirs was a Trumanesque, NSC-68 view of the world. History both tugged at them and pushed them.

The style, personality, and mood of the Kennedy team joined the historical imperatives to compel a vigorous foreign policy. "All at once you had something exciting," recalled Don Ferguson, a student campaigner for Kennedy in Nebraska. "You had a young guy who had kids, and who liked to play football on his front lawn. He was a real human being. He was talking about pumping some new life into the country . . . just giving the whole country a real shakedown and a new image. . . . Everything they did showed that America was alive and active. Family ski trips . . . , Jackie with her new hair styles. . . . To run a country," Ferguson concluded, "it takes more than just mechanics. It takes a psychology."[19] Call it psychology, charisma, charm, image, mystique, or cult, Kennedy had it. He moved Fred Waring and his Pennsylvanians out of the White House and brought Mozart in. Photogenic and quick-witted, he became a media star. Observers marveled at his speed-reading abilities. Decrying softness in the American people, he challenged their egos by launching a physical fitness program.

Handsome, articulate, witty, ingratiating, dynamic, energetic, competitive, athletic, cultured, bright, self-confident, cool, analytical, mathematical, zealous—these were the traits universally ascribed to the young President. People often listened not to what he said but how he said it, and he usually said it with verve and conviction. He simply overwhelmed. "It is extraordinary," W. W. Rostow remarked to a friend, "how the character of the President's personality shapes everything around him. . . . "[20] Dean Rusk remembered Kennedy as an "incandescent man. He was on fire, and he set people around him on fire."[21] For Schlesinger, JFK had "enormous confidence in his own luck," and "everyone around him thought he had the Midas touch and could not lose."[22]

Style and personality are important to how diplomacy is conducted; how we behave obviously affects how others read us and respond to us, and our personal characteristics and needs shape our decisions. Many of his friends have commented that John F. Kennedy was driven by a desire for power, because power ensured winning—that he personalized issues, converting them into tests of will. Everything became a matter of crises and races. His father, Joe Kennedy, demanded excellence. As James Barber has pointed out in his book, *The Presidential Character*, old Joe "pressed his children hard to compete, never to be satisfied with anything but first place. The point was not just to try; the point was to win."[23] John developed a thirst for victory, a self-image as the vigorous man. Aroused in the campaign of 1960 by the stings of anti-Catholic bias, by complaints that he lacked enough experience in foreign affairs to stand up to Khrushchev, by misplaced right-wing charges that he was "soft on Communism," and by his narrow victory over Nixon, Kennedy seemed eager to prove his toughness once in office. "Tough" and "soft" became two of the words most often uttered by the President and his assistants. A "cult of toughness" developed.[24]

Kennedy took up challenges with zest, relishing opportunities to win. Soon Americans watched for box scores on the missile race, the arms race, the space race, and the race for influence in the Third World. Even the program supposedly carrying the least Cold Warriorist character, the Peace Corps, became part of the game. When JFK learned in 1961 that both Ghana and Guinea had requested Peace Corps volunteers, he told Rusk: "If we can successfully crack Ghana and Guinea, Mali may even turn to the West. If so, these would be the first Communist-oriented countries to turn from Moscow to us."[25] Peace Corps Director Sergeant Shriver placed a sign in his office: "Good Guys Don't Win Ball Games."[26] To the missile expert Wernher von Braun, Kennedy wrote: "Do we have a chance of beating [them] by putting a laboratory in space, or by a trip around the moon and back with a man? Is there any space program which promises dramatic results which we could win?"[27] McGeorge Bundy expressed the administration's general frustration over having gained few decisive victories in 1961: "We are like the Harlem Globetrotters, passing forward, behind, sideways, and underneath. But nobody has made a basket yet."[28] They were eager to score and win.

Kennedy and his advisers, it seems, thought Khrushchev and the Russians were testing them as men. In early 1961, when they discussed the possibility of a summit meeting with Khrushchev, Kennedy asserted that "I have to show him that we can be as tough as he is. . . . I'll have to sit down with him, and let him see who he's dealing with."[29] And a White House aide explained the Bay of Pigs invasion of April 1961: "Nobody in the White House wanted to be soft. . . . Everybody wanted to show they were just as daring and bold as everybody else."[30] During those tense hours when the news about the disaster at Cuba's Playa Girón reached the White House, Robert F. Kennedy exploded, claiming that Moscow would think the Kennedy Administration weak unless the mission succeeded. Something had to be done. Walt Rostow quieted him with this counsel: "We would have ample opportunity to prove we were not paper tigers in Berlin, Southeast Asia, and elsewhere."[31] John F. Kennedy and his aides, as somebody has put it, feared to be thought fearful.

With these psychic needs and with their high intellectual talents, the Kennedy officers swept into Washington, "swashbuckling" and suffering from "auto-intoxication," commented one observer.[32] Cocky, thinking themselves the "right" people, they were, complained a skeptical Chester Bowles, "sort of looking for a chance to prove their muscle." They were "full of belligerence."[33] Schlesinger captured the moment this way:

"Euphoria reigned; we thought for a moment that the world was plastic and the future unlimited."[34] In early 1961 the former Harvard historian advised the President on Latin American policy and declared that "the atmosphere is set for miracles."[35] Bustle, zeal, energy, and optimism—along with toughness— became bywords. How upset Kennedy became at a luncheon for Texas publishers when one of them audaciously stood up and said: "Many Texans in the Southwest think that you are riding [your daughter] Caroline's tricycle, instead of being a man on horseback."[36] Soon friendly journalists were called in and asked to help counter this suggestion of a weak President.

The Kennedy people considered themselves "can-do" types, who with rationality and careful calculation could revive an ailing nation and world. Theodore H. White tagged them "the Action Intellectuals."[37] They believed that they could manage affairs, and "management" became one of the catchwords of the time.

With adequate data, and they had an inordinate faith in data, they were certain they could succeed. It seemed everything could be quantified. When a White House assistant attempted to persuade Secretary of Defense Robert McNamara, the "whiz kid" from Ford Motor, that the Vietnam venture was doomed, the efficiency-minded McNamara shot back: "Where is your data? Give me something I can put in the computer. Don't give me your poetry."[38] The problem, of course, was that some of the data on Vietnam was inconclusive or false. "Ah, *les statistiques*," said a Vietnamese general to an American official. "We Vietnamese can give him [McNamara] all he wants. If you want them to go up, they will go up. If you want them to go down, they will go down."[39] With its faith in formulas and the computer, the Kennedy "can-do" team brought a freshness to American foreign policy, if not in substance, at least in slogans: "The Grand Design" for Europe; the "New Africa" policy; "Flexible Response" for the military; the "Alliance for Progress" for Latin America; and the "New Frontier" at home.

The Kennedy style was evident in the President's alarmist Inaugural Address. Its Cold War language was matched by the pompous phrasing that "the torch has been passed to a new generation." He paid homage to historical memories when he noted that that generation had been "tempered by war" and "disciplined by a hard and bitter peace. . . ." Then came those moving, but in hindsight rather frightening words: "Let every nation know that we shall pay any price, bear any burden, meet any hardship, support any friend, oppose any foe to assure the survival and the success of liberty."[40] No halfway measures here. Kennedy and his assistants thought they could lick anything. They were impatient. As Schlesinger recalled, the Kennedy Administration "put a premium on quick, tough, laconic, decided people. . . . "[41] But Ambassador to the United Nations Adlai Stevenson grew disenchanted. As he told a friend: "They've got the damndest bunch of boy commandos running around . . . you ever saw."[42] Robert Kennedy himself later thought about the decisions of his brother's administration. He wondered "if we did not pay a very great price for being more energetic than wise about a lot of things, especially Cuba."[43] Schlesinger later agreed that "the besetting sin of the New Frontier . . . was the addiction of activism."[44]

The Cuban missile crisis provided an opportunity for exercising management skills and for establishing the credibility of containment and the President's toughness. What is most telling about Kennedy's response to the reckless Soviet installation of missiles in Cuba is that he suspended diplomacy and chose a television address, rather than a direct approach to Moscow, to inform Khrushchev that his flagrant intrusion into the Caribbean would not be tolerated. Kennedy practiced public rather than private diplomacy and thereby significantly increased the chances of war. Why?

Ever since the Bay of Pigs muddle, the Kennedy team had nurtured a sense of revenge. In the wake of the invasion's failure, Kennedy blamed the CIA and the Joint Chiefs of Staff for faulty intelligence and execution, but he never questioned his own policy of attempting to overthrow a sovereign government—only his methods for doing so. In his April 20, 1961, speech just a few days after the Cuban tragedy, he seemed emboldened rather than contemplative or contrite. "Let the record show that our restraint is not inexhaustible." He pledged to take up the "relentless struggle" with Communism in "every corner of the globe." Employing once again the theme of toughness vs. softness, the President declared that the "complacent, the self-indulgent, the soft societies are about to be swept away with the debris of history. Only the strong . . . can possibly survive."[45] Privately Kennedy wondered how he could have let such a flawed operation go forward. One foreign policy analyst explained how, in a letter to the journalist Walter Lippmann. Louis Halle wrote from Europe:

> I can imagine how the President got such bad advice from such good advisers. The decision on which they were asked to advise was presented as a choice between action and inaction. . . . None of the President's advisers wants to have it said of him by his colleagues . . . that he . . . loses his nerve when the going gets hot.
>
> The Harvard intellectuals are especially vulnerable, the more so from being new on the scene. They are conscious of the fact that the tough-minded military suspect them of being soft-headed. They have to show that they are he-men too, that they can act as well as lecture. . . . We have learned that, in foreign relations, the ability to wait may be as important as the ability to act.[46]

"The Castro regime is a thorn in the flesh," Senator J. William Fulbright once told the President, "but it is not a dagger in the heart."[47] JFK disagreed. He worked to expunge Fidel Castro and his revolution from the Western Hemisphere. The administration increased CIA aid to Cuban exile groups and commando squads based in Miami, orchestrated the ouster of Cuba from the Organization of American States, and pressed Latin American countries to break diplomatic relations with Havana. The economic blockade was tightened, and America's European allies were lobbied to restrict their trade with the island. The Kennedy Administration also launched Operation Mongoose. This program of "dirty tricks" sought to sabotage property in Cuba and stir up anti-Castro feelings. It destroyed and maimed, but it did not dislodge Castro. And the CIA, told by the Kennedy brothers in vague but emphatic terms to get rid of Castro, continued the plot begun under Eisenhower to assassinate the Cuban leader. "My God," muttered the CIA's Richard Helms, "these Kennedys keep the pressure on about Castro."[48]

McNamara recalled that "we were hysterical about Castro at the time of the Bay of Pigs and thereafter."[49] Had there been no Bay of Pigs, no Mongoose, no sabotage through hit-and-run raids, no campaign to isolate Cuba politically and diplomatically, no assassination attempts—had there been no concerted effort to cripple the Cuban Revolution and murder its leader, the Cuban missile crisis probably would not have occurred. The Russian military build-up on the island in 1962 came in response to United States pressure. Castro may very well have called in the Soviet missiles to deter further American hostile acts, including an American invasion the Cubans so feared. Khrushchev would never had had the opportunity to stage his dangerous missile show if Kennedy had not been working to knock Castro off his perch.

When the missiles were discovered in Cuba in October 1962, Kennedy was poised for boldness, for another test of will, for a cleansing of the Bay of Pigs insult. It was an eyeball-to-eyeball confrontation, and remember, said Rusk, the Russians blinked first. During the crisis, A. A. Berle recorded the following in his diary: "This is reprise on the Bay of Pigs business and this time there will be no charges that somebody weakened at the crucial moment."[50] Kennedy himself put it this way in a letter to Prime Minister Harold Macmillan of Great Britain: "What is essential at this moment of highest test is that Khrushchev should discover that if he is counting on weakness or irresolution, he has miscalculated."[51] Personality and style alone did not determine the American reaction in the Cuban missile crisis. There were obvious strategic calculations. But the *way*, the *manner* in which Kennedy responded was molded by the "action intellectuals'" style and mood. The President's desire to score a victory, to recapture previous losses, and to flex his muscle accentuated the crisis and obstructed diplomacy. Public statements via television are not calculated to defuse a crisis; Kennedy gave Khrushchev little chance to withdraw his mistake or to save face. He left little room for bargaining, but instead issued a public ultimatum. The members of the Executive Committee which advised the President were bright and energetic, Robert Kennedy recalled, but "if any one of half a dozen of them were President the world would have been very likely plunged in a catastrophic war."[52] "We were in luck," John Kenneth Galbraith later commented, "but success in a lottery is no argument for lotteries."[53]

The result? Russia was humiliated publicly. Having its own pride, recognizing its nuclear inferiority, and being harangued by the Chinese as "capitulationist," Moscow launched a massive arms build-up. "Never will we be caught like this again," concluded the Soviet Deputy Foreign Minister.[54] The American lessons for the Cuban missile crisis were, on the other hand, that "force and toughness became enshrined as instruments of policy." Indeed, "the policy of toughness became dogma to such an extent that non-military solutions to political problems were excluded."[55] The 1962 "victory" in Cuba also may have encouraged the Kennedy Administration to take firmer action in Vietnam.

The presidential style, the historical imperatives, and now the third generator of John F. Kennedy's foreign policy: counter-revolutionary thought. Rostow has instructed us that for Kennedy "ideas were tools. He picked them up easily like statistics. . . . He wanted to know how ideas could be put to work."[56] And the intellectuals in the Kennedy court fed the new President a steady diet of ideas. The key concept was "nation-building." Through "modernization" (what Kennedyites called "peaceful revolution"), developing nations would be helped through the stormy times of economic infancy to economic (and hence political) maturity. The Kennedy team understood the force of nationalism in the Third World; rather than flatly opposing it, the "action intellectuals" sought to use it or channel it.

The governing notion was that evolutionary economic development would insure non-Communist political stability. The Alliance for Progress and the Peace Corps sought to fulfill this concept. "Economic growth and political democracy," Kennedy proclaimed, "can develop hand in hand."[57] Washington thus tried to induce what Schlesinger called "middle-class revolution." Land reform, industrialization, tax reform, and public health and sanitation programs had to be undertaken, Schlesinger informed the President, or "new Castros will infallibly arise across the continent."[58] "Modern societies must be built," Rostow told the 1961 graduating class of the Special Warfare

School at Fort Bragg, "and we are prepared to help build them."[59]

Kennedy liked to quote Mao's statement that "guerrillas are like fish, and the people are the water in which fish swim. If the temperature of the water is right, the fish will thrive and multiply."[60] Kennedy sought to affect the temperature of the water through modernization. Counter-insurgency became his chief means. Whether or not Khrushchev had issued his January 1961 proclamation that Russia would support movements of national liberation, the Kennedy team would surely have undertaken counter-insurgency operations. They were committed before Moscow ever uttered its alarming pledge. Insurgencies were destablizing movements, assumed to be Communist-inspired. Bold action was called for to stop the Communist menace.

Counter-insurgency took several forms, all reflecting the "can-do" philosophy. Popular were the training of native police forces and bureaucrats, flood control, and transportation, communications, and community action projects. Most dramatic, and something in which the President took a keen interest, were the American Special Forces units or the Green Berets. Kennedy did not create this elite corps of warriors, but he personally elevated their status and supervised the choice of equipment for them. They would apply America's finest technology in Vietnam to succeed where the French had failed after a ten-year effort. In late March 1961, Rostow urged a "counter-offensive" in Indochina by using "our unexploited counter-guerrilla assets." As Rostow advised JFK, "in Knute Rockne's old phrase, we are not just saving them for the Junior Prom."[61] Kennedy ordered a five-fold increase in the size of the Special Forces and gradually enlarged the number of American military personnel in Vietnam from 685 to more than 16,000 by the end of 1963. As Kennedy's favorite general, Maxwell Taylor, put it, Vietnam became a "laboratory" in counter-insurgency techniques.[62]

The arrogance and bias of these ideas are conspicuous. No matter how one cuts them, they meant significant American interference in the affairs of other nations. They were grand in theory, so pragmatic and humanitarian at the same time. But something went wrong. The nation-building concept simply did not pay proper attention to the world's diversity and complexity, the multitude of indigenous forces, the varied traditions of other cultures, the entrenched position of native elites, and the persuasive appeals of insurgent, nationalistic leftists. It is remarkable how blindly ignorant Rostow and other officials were of foreign cultures when they chose to project the American experience onto others. They ultimately found out that economic growth and democracy do not necessarily go hand-in-hand, that the middle class could be selfish or did not exist at all, that some nations had no tradition of liberal politics, that not all insurgencies were Communist, and that rebels, close to their nation's pulse, believed deeply in their cause and would suffer great sacrifices to gain success. In a revealing question, somebody once asked why "their" Vietnamese fought better than "our" Vietnamese?

The nation-building concept also over-estimated the power of the United States to shape other nations. It assumed that young men and women from Oregon, Iowa, Connecticut, and North Carolina could manage "natives" abroad, much as they had done in the Philippines at the turn of the century or in Latin America through the early twentieth century. But unable to force reform on others, they often ended up violating American principles by supporting the elite or the military or by trying to topple regimes, such as that of Ngo Dinh Diem in Vietnam. Strategic hamlets, part of counterinsurgency in Vietnam, proved to be disruptive of village life. Villagers bitterly resented resettlement. The Vietcong appeared to be Robin Hoods. The concept assumed further that the United States had an obligation to cope with insurgencies everywhere. It made few distinctions between areas key and peripheral to American national interest. It did not define the "threat" carefully; it tried to do too much; it was globalism gone rampant.

The concept also possessed a pro-capitalist, private-enterprise bias, because it favored "private" development. But in the Third World that method was traditionally identified with exploitation, and developing nations were bent upon gaining control over their own natural resources. Finally, nation-building did not estimate the strain that would be placed on American resources and patience in this long-term, global role as policeman, social worker, and teacher. It tended to take for granted the American people and the constitutional system, including congressional prerogatives in policymaking. Overall, then, the revered, clinical concepts of the Kennedy Administration bumped up against a host of realities. Arthur M. Schlesinger, Jr., came to regret the administration's fascination with counter-insurgency—"a mode of warfare for which American were ill-adapted, which nourished an American belief in the capacity and right to intervene in foreign lands, and which was both corrupting in method and futile in effect."[63]

The Kennedy Administration, propelled by Cold War history, by its own bumptious "can-do" style, and by its grandiose theories for world revival, bequeathed a dubious legacy in foreign policy. Would JFK have changed had he lived? It seems unlikely. He would have had to fire the hard-line advisers who persistently clung to their theories. The Cold War was too ingrained in Kennedy's own experience to permit much adjustment. It is not likely that he would have shed his penchant for personalization or boldness. And, like all people who must make decisions they need to justify, he would probably have persisted in defending his mistakes with the distortions necessary.

He temporarily quieted the crisis in Laos and followed a cautious policy in the Congo, and just before his death Kennedy had doubts about Vietnam and about the rigidities of his Cold War stance. His June 1963 address at American University is often cited as the example of his change of heart, for therein he expressed an uneasiness with high weapons expenditures, called for a re-examination of American Cold War attitudes, suggested that conflict with the Soviet Union was not inevitable, and appealed for disarmament. It was a high-minded speech and suggested a willingness to negotiate—one product of which was the Test Ban Treaty. Still, "one speech is not enough," as Kennan has remarked.[64] This speech was not typical of Kennedy or his advisers, many of whom stayed on to work with President Lyndon B. Johnson.

More typical are other elements of the Kennedy legacy: an arms race of massive proportion and fear, reflected in the bomb shelter craze that Kennedy encouraged; a tremendous enlargement of the American nuclear arsenal, even after Kennedy discovered that there was no "missile gap"; neglect of traditional diplomacy, as in the Cuban missile crisis; escalation in

Vietnam; a globalism of overcommitment. Kennedy showed a distinct impatience with the philosophy of "doing little or nothing." As Anthony Hartley has concluded: "The style of the Kennedy diplomacy excluded the attentive watching and patient waiting which are the secret of a successful foreign policy."[65] Kennedy also continued concentration of foreign policy decision-making in the White House and fed the "imperial presidency." Congress was not even informally informed, for example, that the United States intended to attack a sovereign nation at the Bay of Pigs.

The Kennedy team exaggerated the threat posed to the United States by insurgencies and the Soviet Union, clouded distinctions between Communists and insurgents by espousing the "domino theory," and adhered to the simple "zero-sum game" view of world politics that a victory for Communists anywhere represented a loss for the United States. An increasingly critical but still good-humored Galbraith wrote the President in March 1962: "Incidentally, who is the man in your administration who decides what countries are strategic? I would like to have his name and address and ask him what is so important about this real estate in the space age. What strength do we gain from alliance with an incompetent government [in Vietnam] and a people who are so largely indifferent to their own salvation? Some of his decisions puzzle me."[66] The Kennedy Administration, as well, continued a tired, antiquated policy of non-recognition toward the People's Republic of China, even as the Sino-Soviet split widened.

The world was not plastic, the 1960s were not the 1940s, and Kennedy's style of toughness was more appropriate to the football field than to diplomacy. Kennedy did not act like a Cold Warrior simply because he was pressured by recalcitrant right-wingers, a Cold War Congress, or an out-of-control bureaucracy, or because he was bound by inherited—and failed—Eisenhower policies. Kennedy believed in Cold War dogmas himself and gave them a new vigor. He was thus not only a maker of history but a victim of it. Arrogance, ignorance, and impatience, combined with the familiar exaggerations of Soviet capabilities and intentions, helped make the world a more dangerous place than when he took office.

# 11

## Questioning the Vietnam War: Isolationism Revisited

Hawks and doves alike have reckoned the costs and consequences of Vietnam—that prolonged war that seemed to have
no precise beginning and no end until 1975, when Americans
were driven abruptly and ignobly from their embassy in Saigon.
By official count, over 58,000 Americans died in the conflict.
Well over a million people in the small countries of Cambodia,
Laos, and Vietnam perished. The United States spent some $200
billion to wage war in Southeast Asia, and will pay out close to
that amount in future veterans' benefits. The tremendous war
expenditures fueled inflation at home, wrenching the economy.
And there was what economists call "opportunity cost"—
potential gains never realized because resources were invested
in the war. The civil rights movement and other domestic
reform efforts, including President Lyndon B. Johnson's Great
Society, became crippled as the war increasingly absorbed funds
and attentions. American politics became wounded too: Americans became disaffected, distrustful of leaders who, it seemed,
lied to them—and would again. The political storms churned up
by Vietnam persuaded Johnson to shun running for another
term, and in 1968 helped defeat Hubert H. Humphrey and elect
Richard M. Nixon. Watergate had Vietnam sources too: Indulging a seige mentality because of growing opposition to the war

and vowing to stop leaks of war information damaging to the government, White House aides dispatched a law-flouting "plumbers" group to intimidate critics. They and their President got caught in a host of dirty tricks that ultimately forced Nixon into resignation to avoid certain impeachment. America's longest war also alienated allies in Europe and widened the chasm between the United States and Third World nations. Vietnam delayed détente and nuclear arms control agreements. And Vietnam weakened the American military by slowing its modernization, creating dissent within its ranks, and bringing doubt upon its ability to fight and win. In the end, a war its Cold War architects believed necessary to American national security damaged that security. One American diplomat wondered "how so many with so much could achieve so little for so long against so few."[1]

As they suffered with others the costs and consequences of Vietnam in the late 1960s and early 1970s, critics protested that the price was too high. They asked whether the threat was so great that it demanded such sacrifice, such damage not only in Vietnam but at home. Pointing to civil war rather than Communist aggression as the root of Southeast Asian turmoil, they claimed that the containment doctrine tenaciously sent forward from the Truman years as inherited wisdom was being misapplied—that Ho Chi Minh was an Asian Tito whose nationalism prevented him and his Vietnamese followers from ever becoming puppets of either Soviet Russia or Communist China. But, more, containment had become global anti-Communism, inducing overcommitment and indiscriminate interventionism, in themselves as much a threat to the United States as dreaded Communism. When dissenters asked who or what the United States was containing in Vietnam, they received, at different times and with different emphases, a multitude of answers: the Soviet Union, Red China, Ho's North Vietnam, the South's Vietcong, wars of national liberation, and, of course always, Communism. But with the Sino-Soviet split gaping for all to see and the myth of monolithic, international Communism exposed and discredited, "Communism" would not wash as a convincing explanation.

In their debate with the Johnson and Nixon Administrations, Vietnam War critics looked to the past for guidance; they sought ways to use history to question official Washington's case. They rediscovered the views of early Cold War critics like Henry A. Wallace and Robert A. Taft (see Chapter 6). They studied the anti-war statements of Robert LaFollette and Jeannette Rankin, both of whom had challenged President Woodrow Wilson's decision for American entrance into the First World War. They also delved into the history of anti-imperialism, finding some intellectual compatibility with the 1890s anti-imperialist arguments of William Jennings Bryan, Mark Twain, Jane Addams, Samuel Gompers, and William James. And, as well, they reconsidered the non-interventionism of the isolationists of the 1930s.

As Vietnam cut its deadly path through American life, the word "isolationism" resurfaced. Some critics, like Walter Lippmann and George F. Kennan (see Chapter 7), allowed that they had become comfortable with the label. "Compared to people who thought they could run the universe," remarked Lippmann, "I *am* a neo-isolationist and proud of it."[2] The political scientist Robert Tucker wrote a book in 1972, titled *A New Isolationism*, which defended the idea. Bruce M. Russett, another prominent political scientist, published *No Clear and Present Danger* that same year. The book's subtitle revealed his purpose: "A Skeptical View of the U.S. Entry into World War II." Others, like Senator J. William Fulbright, found in the isolationists of the depression decade a critique of American foreign policy that informed a critique of Vietnam.

The war's managers, on the other hand, revived the term "isolationism" to heap derision on their opponents. In a 1969 speech, President Nixon branded critics of his foreign policy "new isolationists," and suggested that they were turning their backs on the world.[3] It was a phrase meant to belittle and discredit; it conjured up the specter of a 1930s United States shirking international responsibility and thus helping to unleash mad dictators and insatiable militarists. The label further summoned the memory of America's rise to world eminence after World War II, throwing off its isolationism and assuming its rightful first rank in the international community. The pejorative tag "isolationism," then, connoted weakness, appeasement, surrender, mindless idealism, and a retreat from the obligations of world leadership.

Scholars too had dealt critically with the non-interventionists of the 1930s. Historians had complained that the isolationists ignored the threat of totalitarianism, left the United States militarily unprepared, courted fascism, and in general nurtured a wrongheadedness that restrained America from influencing international events on the road to World War II. Typical was the conclusion that "Congressional isolationists, so anxious to keep out of war, actually helped invite a foreign catastrophe of such immense proportions that no nation could have escaped its consequences."[4] Although the isolationist movement was notably diverse—Nazis, Communists, pacifists, defeatists like Charles Lindbergh, "fortress America" types like Herbert Hoover, anti-Semites like Father Charles E. Coughlin, liberal reformers like George Norris, and scads of intellectuals at one time or another associated with it—isolationism has often been linked to conservatism. This is so because the America First Committee, the most conspicuous isolationist organization, housed some unprogressive businessmen, and because many isolationists, especially in the late 1930s, opposed the New Deal. Actually, many liberals quit the isolationist camp in 1939–41.

There was much in isolationism that Vietnam critics, reading its history, could not admire. The Ludlow Amendment, which would have required a public referendum on all war decisions, was ill-advised. The isolationists sometimes imagined war plots and conspiracies. The isolationist or "revisionist" histories of the coming of World War I written by Charles Tansill and others were less than scholarly and sharply polemical. Above all else, the isolationists were wrong in the 1930s to believe that the United States could avoid involvement in the developing crises in Europe through the Neutrality Acts, which prohibited America from punishing the aggressor.

But critics found that the isolationists had provided a long-term and constructive critique of American foreign policy. Even in their failure the isolationists left an enduring analysis of facets of United States diplomacy. Those isolationists who were domestic liberal reformers especially earned attention, for their assessment of foreign policy was the best articulated, the least rigid, and the most meaningful for the 1960s and the 1970s. Conservative isolationists did not dominate isolationism, and Manfred Jonas, in his historical study of isolationism, suggested that isolationists be accorded intellectual respect: "To regard isolationism as pure obstructionism . . . is unfruitful and misleading." Gerald P. Nye, Norman Thomas, Charles Beard, John Bassett Moore, and William Borah, among others, "did not approach American foreign policy from a purely negative and obstructionist viewpoint." Isolationism "was the considered response to foreign and domestic developments of a large, responsible, and respectable segment of the American people."[5]

In the 1930s, indeed, the term "isolationist" was not a label of reproach, but rather one respectably worn by many Americans, perhaps including even President Franklin D. Roosevelt. Isolationism did not mean isolation. Most isolationists desired foreign trade, continued immigration and cultural exchange, and diplomatic intercourse with other nations. Senator Borah, for example, favored international steps to promote disarmament and in 1933 led the fight for American recognition of the Soviet Union. Nor was isolationism the phenomenon of one region, one political party, one ethnic group, or one socioeconomic stratum. It was national, and had three basic strains of thought which attempted to answer the key question: isolation from what?

The first strain was a profound abhorrence of war, a fear of war and militarism. Isolationists recoiled not only from the bloodshed of war and the waste of total mobilization, but also from the detrimental effects of war on domestic reform and civil liberties. The second ingredient was non-intervention—that the United States should avoid interference in the affairs of other nations, should be selective in its foreign involvement, and should disengage from empire. The third strain was unilateralism or independence. That is, the United States should preserve its freedom of choice in foreign relations by avoiding binding and restrictive alliances or commitments.

With these three core ideas, the isolationists dissected, studied, and influenced American foreign policy in the 1930s. Their assessment showed them to be aware of some of the realities of world politics and capable of compromise. Many of them, too, changed their minds in the last few years before American entry into World War II. The three ideas of freedom of action, non-intervention, and abhorrence of war later carried relevance for dissenters from the Vietnam War.

The isolationists intensely feared war. The National Farmers Union in 1936 argued that "war is an utter negation of civiliza-tion. It is a relic of barbarism, regimenting mankind in organized murder, starvation, disease, and destruction. It is incompatible with every moral and Christian teaching."[6] Congressman Louis Ludlow declared that the "science of slaughter has advanced with vast strides during the last twenty years . . . The art of killing people en masse and maiming and wrecking human bodies has been perfected until it is impossible to imagine the next large-scale war being anything less than a vast carnival of death. . . . "[7]

Many isolationists believed that the United States had been drawn unnecessarily into World War I, that civil liberties had been blatantly curtailed at home, and that the war was not the noble crusade President Woodrow Wilson claimed it was. The isolationists of the 1930s accepted measured appropriations for national defense, but citing the experience of the First World War, they tried to twinge the American conscience to the horrors of war and the futility of unchecked military escalation. "We tried the madman's way of helping last time," wrote the Socialist leader Norman Thomas, "and we added to the ruin."[8]

Isolationists with a commitment to domestic reform were part of a tradition of opposition to intervention and war in favor of attention to domestic priorities. Norman Thomas asked the question which troubled some Americans after World War I: "If we go into the [second] war, what will happen to the things we prize? What will happen to decency and tolerance, to morality and culture, to democracy and civil liberty? . . . [9] Thomas agreed with the historian Charles Beard and several congressmen who suggested that America put its own house in order to stand as an example of social progress to the world. Only a viable social democracy could combat the evils of fascism. Senator Nye called for "correcting our own ills . . . saving our own democracy rather than soliciting the trouble to come from any move to police and doctor the world".[10]

Liberal isolationists emphasized the social problems created by the Great Depression, and envisioned the death of the New Deal, repression of civil liberties, and "domestic despotism."[11] "If any person can see hopes for democracy in another military and naval crusade for democracy, after looking at the fruits of the last crusade," Beard wrote in 1938, "then his mind passeth my understanding." Business might enjoy profits from a new war, but "the probabilities are that we should then have universal fascism rather than universal democracy."[12] The most pessimistic of all, perhaps, was Stuart Chase, who saw "the liquidation of political democracy, of Congress, the Supreme Court, private enterprise, the banks, free press and free speech; the persecution of German-Americans and Irish-Americans, witch hunts, forced labor, fixed prices, rationing, astronomical debts, and the rest."[13] Although Chase's forecast was obviously exaggerated, in twentieth-century American history war has often undermined domestic reform movements and created an atmosphere conducive to abridgment of civil liberties. Then, too, the isolationists were witnessing the shenanigans of the Un-American Activities Committee of Martin Dies, who condemned people for their ideas, not for their actions. Nye, opposing conformity of thought, reminded Americans that dissent did not mean "disloyalty."[14]

Believing in non-intervention, many isolationists catalogued the long history of United States intervention in Latin America. In 1930, the United States had troops stationed in Nicaragua and Haiti, and held protectorates over the Dominican Republic, Cuba, and Panama. The anti-imperialist isolationists questioned official Washington's assertion that the United States was backing "democracy" in Latin America. At least twelve of the twenty Latin American nations were ruled in 1938 by dictators supported by Washington. "It would seem," concluded Beard, "that the rhetoric of democratic solidarity in this hemisphere does not get very far below the surface of things."[15] Nye condemned the intervention of the marines in Nicaragua in 1927 and their presence there until 1933 "to crucify a people . . .

merely because Americans have gone there with dollars to invest and we have made it our policy to give whatever protection is possible to those dollars."[16] Others complained that the Good Neighbor Policy amounted to little more than window dressing.

Isolationists also opposed unlimited arms sales abroad. John Wiltz, a scholar who has studied the isolationist congressional investigating committee headed by Nye, has written that the committee "proved . . . that most munitions sales to Latin America, China, and the Near East depended upon bribery. . . . It established that munitions firms sometimes played one belligerent off against another. . . . The Committee exhibited documents which shocked many Americans into the realization that there was a difference between selling instruments of human destruction and selling sewing machines and automobiles."[17] With the encouragement of isolationists, Roosevelt banned all arms shipments to Bolivia and Paraguay in the Chaco War. The isolationists may also have stimulated the partial United States retrenchment from Latin America under Roosevelt. The Nye Committee revealed questionable activity on the part of American companies in international cartels, but did not prove the claim of some isolationists that American business had drawn the United States into World War I.

In pointing out that the United States itself often acted like an imperialist, the isolationists also inspected quite closely the foreign relations of one of America's "natural allies," Great Britain. Believing that all nations act in self-interest, that no nation has a monopoly on virtue and morality, the isolationists asked Americans to stop applying an international double standard: one set of rules for themselves and the British and another for other nations. Robert Hutchins, the young Chancellor of the University of Chicago, fixed on realities: "Mr. Roosevelt tells us we are to save the democracies. The democracies are presumably England, China, Greece, and possibly Turkey. Turkey is a dictatorship. Greece is a dictatorship. China is a dictatorship." He went on: "And what do we do about countries which were victims of aggression before 1939? . . . What do we do about Hong Kong, the Malay States, the Dutch East Indies, French Indo-China, Africa, and above all, India?"[18] Isolationists compared Italy's subjugation of Ethiopia to Britain's role in India.

Isolationists also examined instances of business diplomacy in Latin America and protested the sale of arms to Japan and Germany in the 1930s. Such munitions businessmen were called "merchants of death," who particpated in "rotten commercialism"—"inhumane, immoral, and un-Christian."[19] In part this view sprang from the reform zeal of isolationists who wanted to abolish child labor, improve factory safety conditions, pass anti-trust legislation, curtail monopolies and the misuse of America's natural resources, and lift the living standards of the disadvantaged. They logically asked: If we are fighting big business at home, why help to extend its power overseas? Those parts of the Neutrality Acts which curtailed loans and arms shipments to belligerents, then, reflected the reform movement of the 1930s, much as the Vietnam era anti-war movement drew strength from domestic reformers.

Senate investigations during World War II demonstrated that the isolationists could claim validity for their suspicions that some businesses were compromising the national interest for profit in the 1930s. The American business press was overwhelmingly anti-fascist, but twenty-six of the top one-hundred American corporations of 1937 were linked through significant cartel and contractual agreements with Nazi Germany. And fifty-six American companies were connected with the backbone of Hitler's war machine, the I. G. Farben Company. Standard Oil helped Germany develop both synthetic rubber and 100-octane aviation fuel, but would not develop the latter for the American Army because of a restrictive agreement with Farben. Bendix Aviation, controlled by General Motors, provided a German company (Robert Bosch) in 1940 with complete data on aircraft and diesel engine starters in return for royalties. As an official of the Dow Chemical Company indiscreetly remarked in the 1930s: "We do not inquire into the uses of the products. We are interested in selling them."[20] Some businessmen continued to ship scrap metal and oil to Italy, even though the President asked for a voluntary embargo in 1935. In fact, in the last three months of that year, American oil shipments to Mussolini's Italy tripled. The isolationists proved important watchdogs over the foreign entanglements of American business, alerting Americans to consider the proposition that the national interest and economic self-interest could be at odds.

Some critics of the isolationists have incorrectly interpreted isolationism to mean friendliness toward fascism, or perhaps a condoning of fascism. The isolationist camp did attract fascists, but they hardly constituted the movement. Indeed, numerous liberal isolationists were early and vociferous in their denunciation of Hitlerism and the persecution of the Jews. Oswald Garrison Villard, for example, in 1933 and 1935 urged Western Europe to boycott Hitler's Germany and Mussolini's Italy, and in 1936 appealed to the League of Nations to act against Hitler. Villard became despondent, as did many isolationists, when the other European nations themselves would do nothing to contain the dictators. "The overwhelming majority of all isolationists," Jonas has written, "had no desire to see the Axis Powers gain their ends."[21]

Isolationists certainly should not be held responsible for the coming of the Second World War. Germany did not depend upon American isolationism in making its plans. Britain and France let Germany nibble at vulnerable lands for years, conceding to fascist demands. Not until April of 1939 did France and Britain guarantee the independence of Poland. The League of Nations was moribund in the 1930s, its members unwilling to take decisive action. The Soviet Union was excluded from membership until 1934 and then ousted in 1939. Germany was admitted to the League in 1926 and departed in 1934. Both Japan and Italy withdrew in the mid-thirties. Britain was more interested in balance of power and its interests than in collective security. European appeasers, then, co-authored the American Neutrality Acts, because some Americans sensibly concluded at the time that they had better steer clear of the chaos in Europe if the Europeans themselves could not remedy it.

Charles Lindbergh summarized in 1941 an oft-heard isolationist position that critics of the Vietnam War later found attractive:

> Democracy is not a quality that can be imposed by war. The attempt to do so has always met with failure. Democracy can spring only from within a nation itself, only from the hearts

and minds of the people. It can be spread abroad by example, but never by force. The strength of a democarcy lies in the satisfaction of its own people. Its influence lies in making others *wish* to copy it. If we cannot make other nations *wish* to copy our American system of government, we cannot force them to copy it by going to war.[22]

Vietnam, argued its critics who used the story of the isolationists to make their case, was not only a doomed attempt to spread an American model to an inhospitable environment; it was also a destructive force at home—all in the elusive but traditional quest to meet an ill-defined Communist threat.

# 12

## Nixon, Kissinger, and Détente: New Lever of Containment

"I know that my reputation is one of being a very hard-line, cold war-oriented, anti-communist," President Richard M. Nixon told President Leonid Brezhnev in May 1972 at the Moscow Summit. But, now, Nixon assured the Soviet leader, he believed capitalism and Communism could "live together and work together."[1] Nixon always relished the "big play"—a spectacular diplomatic turnaround to boost his fame.[2] And he liked to be known as a man who would "turn like a cobra" on those who dared to cross him.[3] His self-professed "madman theory" would supposedly deter America's adversaries, who would shy away from triggering his reputed irrationality.[4] Rational or irrational, this secretive, suspicious President decided to move American foreign policy from containment through confrontation to containment through negotiation. Détente became the new lever for this strategy of meeting the Communist threat.

Détente meant limited cooperation, but continued rivalry with the two chief Communist states, the Soviet Union and China. Détente was a means or process intended to produce an international "equilibrium" or balance of power by containing both nations, as well as curbing revolution.[5] Nixon and his chief adviser, Henry A. Kissinger, attempted to exploit the Sino-Soviet split, to pit the two Communist countries against one another, to keep one worrying about what the United States was doing with the other. Thus the two could contain one another. "The President is in the position of the lovely maiden courted by two ardent swains [China and Russia], each of whom is aware of the other but each of whom is uncertain of what happens when the young lady is alone with his rival," wrote a journalist during Nixon's dramatic February 1972 trip to the People's Republic of China.[6]

Détente promised advantages. Moscow and Beijing might help the United States extricate itself from Vietnam. The Cold War was costing too much; détente might reduce the necessity for interventions, spiraling military expenditures, and new nuclear weapons. By the early 1970s, the Soviet Union had achieved nuclear parity with the United States; détente might prevent the Soviets from overtaking Americans. Beijing also possessed nuclear capability, but it had refused to sign international atomic anti-proliferation controls. Without Chinese adherence, nuclear arms control agreements might prove feeble, and it was hoped that détente would bring the Chinese into line. Economic benefits were also calculated. Increased trade with and investment in the two Communist nations might smooth relations to permit the solution of other questions, and the sagging American economy surely would be stimulated. The opening to China would likely thwart reconciliation between Moscow and Beijing as America played its "China card," thus tying down several Soviet military divisions in Asia—away from NATO. And it would improve Asian stability by nurturing new ties between Japan and China. Détente also would enhance Nixon's reputation and prove politically valuable in the 1972 presidential campaign.

In the pursuit of détente, Nixon and Kissinger enjoyed successes. Abandoning a policy of containment through isolation toward the People's Republic of China, they orchestrated the President's surprising visit to Mao Zedong and Zhou Enlai. As the Chinese and Americans toasted one another in the Great Hall of the People on February 21, 1972, Nixon quoted Mao himself: "Seize the day, seize the hour." The euphoric President remarked, "this is the hour."[7] Americans stopped saying "Red Chinese"; Chinese stopped saying "Running Dogs of Imperialism." The business at hand was the containment of the Soviet threat both nations identified. For the Chinese, their 4,150 mile border with the hostile Soviet Union remained a constant source of danger; military skirmishes like those in 1969 might erupt again. Revitalizing the old Chinese practice of using barbarians to control barbarians, the Chinese reasoned that improved Sino-American relations might help deter the Soviets. Although they differed on the status of the other China—Taiwan—and Vietnam, the conferees agreed that neither the United States nor the People's Republic of China "should seek hegemony in the Asia-Pacific region and each is opposed to efforts by any other country or group of countries to establish such hegemony"—a thinly veiled slap at Russia.[8]

In 1973 Washington and Beijing exchanged "Liaison Offices" or mini-embassies, but formal diplomatic relations would wait until 1979, after Watergate, Nixon's resignation, the 1976 presidential election, the deaths of Mao and Zhou, and new political alignments within China. Although the most expansive economic dreams were seldom realized, companies like Boeing, Monsanto Chemical, and Radio Corporation of America signed contracts with the Chinese. Beginning in early 1981 one of America's well-known symbols, Coca-Cola, was bottled in China, where it was known as "tasty happiness." Chinese-American trade began to climb. In 1977–80 China ranked fourth in the world as a buyer of United States agricultural exports, taking about half of the nation's cotton exports in 1979 alone. By 1980 total American sales to China hit $4 billion, up from $700 million in 1973. China replaced Russia as the United States' largest Communist trading partner. By comparison, American exports to the Soviet Union in 1980 stood lower at $1.5 billion.

While cultivating China, the United States also launched negotiations with China's nemesis, Russia. Nixon and Kissinger journeyed in May 1972 to Moscow, where they struck agreements with the Soviets on cooperation in space exploration (a joint space venture was launched in 1975) and trade (large grain sales soon followed). The leaders also discussed Vietnam and concluded that small nations should not interfere with détente. Only a few weeks earlier, when Nixon had escalated the bombing of North Vietnam, he feared that an angry Moscow might cancel the summit. The Russians did not; to them détente came first.

The summit conferees concentrated on the Strategic Arms Limitation Talks (SALT) agreements. When the Nixon Administration took office, it inherited a legacy of doctrines and missiles that defined United States nuclear strategy. In the 1960s, the doctrine of "massive retaliation" evolved into the concept of "mutual assured destruction" or MAD. That is, it was assumed that each adversary was capable of inflicting such heavy losses on the other's industry and population that neither would launch a nuclear attack upon the other. MAD's viability depended upon each side's "second-strike capability": the capacity to absorb a first strike and still destroy the attacker with a retaliatory or second strike. By 1969 American strategists sought a superiority of forces through what was called the triad: land-based intercontinental ballistic missiles (ICBMs), long-range B-52 bombers, and submarine-launched ballistic missiles (SLBMs), all armed with nuclear weapons. To help guarantee superiority, the United States had also begun to flight-test the "multiple independently-targetable reentry vehicle" (MIRV), a vehicle equipped with a warhead and mounted with several similar vehicles on one ballistic missile. Once separated from the missile, each MIRV can be directed against a different target. Finally, President Nixon also inherited initial planning for an "anti-ballistic missile" (ABM) system to defend cities and ICBMs thought vulnerable to Soviet attack. ABMs were designed to intercept and neutralize incoming warheads. Because ABMs theoretically—their efficiency was questionable—protected offensive weapons from attack, critics feared that the other side would be encouraged to build more missiles to overcome the ABM protection, thus further stimulating an already accelerated nuclear arms race.

By 1968 the United States had deployed 1,054 ICBMs to the Soviets' 858; the United States led in SLBMs, 656 to 121, and in long-range bombers 545 to 155. The United States also stood superior in total nuclear warheads, about 4200 to 1100, and in the accuracy of its weapons systems. Yet American officials knew that the Soviets were constructing new missiles, submarines, and bombers at a faster pace. In a few years these new weapons would give the Soviets nuclear parity with the United States. When Nixon and Kissinger began to direct American foreign policy, the two great nuclear powers had become "fencers on a tightrope: each facing the other, weapon in hand, balancing precariously; neither willing to drop his weapon and give way to the other; each fearing to thrust decisively because such a thrust would topple them both, attacker and victim, to mutual disaster."[9]

President Nixon soon abandoned the no longer tenable doctrine of superiority and accepted "sufficiency," or parity of forces with the Soviet Union.[10] Still, he decided to phase in the ABM system, for which Congress, after heated debate and a close vote, provided funds. Nixon also ordered the installation of MIRVs. Thus, the United States could enter the SALT talks, he said, from a position of strength.

The first SALT talks began in Helsinki in November 1969 and alternated between that city and Vienna until 1972. SALT-I culminated on May 26, 1972, at the Moscow summit with the signing of two agreements. The first, a treaty, limited the deployment of ABMs for each nation to two sites only. In essence the accord satisfied the MAD doctrine, because it left urban centers in both countries vulnerable. The other accord, an interim agreement on strategic offensive arms, froze the existing number of ICBMs already deployed or in construction. At the time, the Soviet Union led 1,607 to 1,054. The interim agreement also froze SLBMs at 740 for the USSR and 656 for the United States, although the two nations were permitted to raise these numbers to 950 and 710 respectively if they dismantled one ICBM for every SLBM added. SALT-I did not limit the hydra-headed MIRVs, thus leaving the United States superior in deliverable warheads, 5700 to 2500. Nor did the agreement restrict long-range bombers, in which the United States ranked first with about 450 bombers compared to about 200 bombers for the Soviets. Finally, SALT-I did not prohibit the development of new weapons. The United States, for example, moved ahead on the Trident submarine (to replace the Polaris-Poseidon fleet), the B-1 bomber (to replace the B-52), and the cruise missile (a highly accurate, low-flying guided missile). Indeed, as Kissinger told the defense secretary: "The way to use this freeze is for us to catch up."[11] Hearing that kind of reasoning and witnessing persistently high defense budgets, some analysts, like Herbert Scoville, Jr., former deputy director of the CIA, concluded that "arms control negotiations are rapidly becoming the best excuse for escalation rather than toning down the arms race."[12] Still, SALT-I marked an unprecedented step in advancing frank strategic arms talks and in placing limits on specified nuclear weapons. In August 1972 the Senate passed the ABM treaty by an 88-2 vote; a joint congressional resolution later endorsed the interim agreement. Détente's reputation soared; it obviously had not checked the nuclear arms race, but it had provided a negotiating environment that would also produce SALT-II in 1979.

In Europe détente also worked to ease tensions. Willy Brandt, the West German Chancellor, pursued a policy of *Ostpolitik* to remove the two Germanies from great power competition. A West German-Soviet treaty of August 1970 identified détente as the goal of both countries and recognized the existence of two Germanies. A few months later Brandt signed an agreement with Poland that confirmed the latter's postwar absorption of German territory to the Oder-Neisse line. Then in June 1972 the four powers occupying Berlin signed an agreement wherein Russia guaranteed Western access to the city and relaxed restrictions on travel between the two Berlins. Finally, in December 1972 the two Germanies themselves initialed a treaty that provided for the exchange of diplomatic representatives and membership in the United Nations for both. The Berlin Wall still stood as a reminder of a bitter Cold War past, but it had been hurdled.

The Nixon Doctrine, announced in July 1969, declared that henceforth the United States would supply military and economic assistance but not manpower to help nations defend themselves. This aspect of détente meant, apparently, that the United States was retiring its badge as the world's policeman. It was not that simple. Third World countries held a place in the Nixon-Kissinger scheme for equilibrium because of their vulnerability to radicalism and hence to pernicious Soviet influence. American leaders cited Moscow's endorsement of national liberation movements to argue, therefore, that the internal politics of developing nations were intertwined with the international struggle against Communism. When troubles arose in the Third World, Washington's first impulse was to interpret them as moves in the game of great power politics.

Problems in the Middle East were interpreted in that way, and the tensions there sorely tested détente. By the time Nixon and Kissinger entered office in early 1969, the Middle East, said the President, had become a "powder keg."[13] The Nixon Administration worried that the persistent Arab-Israeli conflict would give the Soviets an avenue into the Middle East. In the spirit of détente, Washington and Moscow began talks on the Middle East that proved futile. Egypt insisted on Israeli withdrawal from occupied territory; Israel steadfastly refused to leave. As American Phantom jets began to arrive in Israel, as Washington wrestled with a new Israeli request for many more aircraft and tanks, and as the Israelis conducted bombing raids deep into Egypt in January 1970, the Soviets began their own military escalation. In the spring of 1970, Soviet surface-to-air missiles (SAMs) were shipped to Egypt to provide a defense system against Phantom attacks. Thousands of Soviet troops, advisers, and pilots answered Egypt's call for assistance. Washington growled at Moscow and gave Israel more F-4s and electronic equipment to improve Israeli accuracy against Egypt's missile sites. "Was this the stuff of détente?" people asked.

Soviet relations with the Palestine Liberation Organization (PLO) and Egypt grew rocky in the early 1970s, as Moscow tried to restrain both of them out of fear that Washington would scuttle détente if Middle East tensions continued. For its part, the Nixon Administration from 1970 to 1973 followed a "standstill diplomacy."[14] It assumed that the Middle East had been stabilized by the sustained arming of Israel ($1.2 billion in American military credits, 1971–73), Soviet restraint, and the emergence of a seemingly more moderate Egyptian government under Anwar al-Sadat (after Nasser died in September 1970). Actually, Sadat was plotting a new war against Israel. He asked the Soviets for more weapons, but received few. In the summer of 1972 he abruptly expelled several thousand Soviet technicians and military advisers. Then, on October 6, 1973, Egyptian and Syrian forces struck Israel. The attack took Israel and the United

States by surprise. At first the Israelis suffered heavy losses, and the Arabs regained land lost in 1967. Nixon promised more Phantoms. "We will not let Israel go down the tubes," he said.[15] Moscow responded by transporting military equipment to Syria.

In the midst of the crisis, the shadow of Watergate lengthened over the Nixon Administration. On October 10, Vice President Spiro Agnew resigned after evidence surfaced that he had accepted payoffs as governor of Maryland years before. Ten days later Nixon fired the special Watergate prosecutor for getting too close to damaging evidence. The coincidence of domestic and foreign crises frayed nerves. The White House staff feared that Moscow might think the government was weak or incapacitated.

On October 13 Nixon ordered a massive airlift of military material to Israel. Soviet Premier Aleksei Kosygin flew to Cairo in the hope of persuading Sadat to accept a ceasefire. Suspicious, Kissinger advised that the Soviets be "run into the ground" by delivery of more American equipment to Israel than the USSR could match.[16] Kissinger himself flew to Moscow on October 20, learning en route that the Saudis had embargoed oil shipments to the United States. By October 21 most of the Arab members of the Organization of Petroleum Exporting Countries (OPEC) had joined the embargo. Kissinger and the Soviets finally arranged a ceasefire in the Mideast war on October 22. But the Israelis, who were now winning, violated the truce lines. Moscow then angrily threatened intervention, and both Washington and Moscow put their forces on alert. Kissinger admonished the Israelis to honor the ceasefire. This time they did, and a new ceasefire held.

The Arab-Israeli contest threatened the American economy. Arab states like Saudi Arabia, which for three decades had supplied Western nations with inexpensive petroleum, were now using their black riches as a weapon: They placed an embargo on petroleum shipments to the United States and quadrupled the price of crude oil for Western Europe and Japan. The United States, importing between 10 and 15 percent of its oil from the Middle East, endured an energy crisis. Gasoline prices at the pumps spun upward and anxious drivers lined up, sometimes for hours, hoping to fuel their automobiles. The embargo was lifted in March 1974, but prices remained high and America's vulnerability had been exposed. Kissinger launched "shuttle diplomacy" to prevent another Middle East blow-up. With impressive stamina and patience, he bargained in Cairo and Tel Aviv intermittently for two years. President Sadat was soon saying, "Dr. Henry, you are my favorite magician."[17] Washington also pressed Israel. Finally, on September 1, 1975, Egypt and Israel initialed a Kissinger-designed agreement that provided for an eventual Israeli pullback from part of the Sinai, created a United Nations-patrolled buffer zone, and placed American technicians in "early warning" stations to detect military activities. Washington also tendered promises of substantial foreign aid to both Egypt and Israel.

Thorny problems remained in the Middle East. The Palestinian Arabs still lived in refugee camps and demanded a homeland, while Israelis entrenched themselves in occupied territories, building industries, farms, and houses. Jordan still demanded the return of the West Bank, and Syrian-Israeli hostility persisted as the Golan Heights remained in Israel's hands. Moreover, a bloody civil war broke out in Lebanon, which prompted Syria to send in troops in 1976. Sophisticated American weapons continued to be shipped to both the Arabs and Israelis after the October war, and Sadat warned that military conflict could erupt again. Egypt's economy remained unstable, fanning political unrest. In March 1976, Sadat, who needed American technology and mediation, denounced Russia, saying that "99 percent of the cards in the game are in America's hands whether the Soviet Union likes it or not."[18] Once Cairo turned so emphatically toward the United States, American policy in the Middle East looked more like old-fashioned containment than détente.

As a counterweight to "Soviet intrusion and radical momentum" in the Middle East, the Nixon and Ford Administrations fashioned a closer alliance with the Shah of Iran.[19] Nixon and Kissinger visited Teheran in 1972 and promised the Shah all the non-nuclear weapons he wanted and American technicians to help Iranians operate the sophisticated hardware. The Iranian military gorged itself on huge amounts of modern American arms that the police-state monarch bought, thanks to galloping oil revenues ("petro-dollars"). American corporate executives rushed to Iran to display their submarines, fighter aircraft, assault helicopters, and missiles. The Shah bought generously. In 1977 his nation ranked as the largest foreign buyer of American-made arms, spending $5.7 billion that year alone. For 1973–78 the bill was $19 billion. His armed forces became the most powerful in the region. But doubters in the shahdom thought such excessive military spending foolhardy when the Iranian per capita income was only $350 and such funds could be better applied to alleviating the nation's economic woes. Also, the Shah's CIA-trained secret service jailed or killed critics with a ruthlessness that drew protest from many Americans. To improve his image in the United States he hired a New York advertising agency and lavishly bestowed gifts on prominent Americans. Official Washington regarded Iran as a pillar of stability in the turbulent Middle East, but critics argued that the huge infusion of weapons actually caused instability.

If détente proved fragile in the Middle East, it also sputtered elsewhere as a vehicle for the reduction of great power conflict and the settlement of crises. In Africa, the United States and the Soviet Union tangled in Angola, where they supported opposing factions in that new nation's civil war. When in 1975 the Soviet-backed group won, Washington denounced Moscow as an aggressor (see Chapter 13). In Latin America, the United States and Soviet ally Cuba took a few halting steps toward a relaxation of tensions, but when Cuban troops helped the Angolan leftists gain power, American officials denounced Havana as a Soviet surrogate. In Chile, Nixon and Kissinger saw the Soviets where they were not. In 1973, using the CIA, they helped topple the elected, leftist government of Salvador Allende (see Chapter 13) in a classic case of exaggerating the Communist threat. The shaky international economy also nettled détente, destabilizing the order détente was supposed to emplace. A worldwide recession, inflation, elevating oil prices, debt-ridden, food-poor Third World nations rife with internal conflict, a North-South debate over tariffs, natural resources, multinational corporations, and interest rates, and the breakdown of the international monetary system—all unsettled the diplomatic envirnoment. "History has shown that international political stability requires international economic stability," Kissinger said in the peace and prosperity idiom so popular some thirty years before.[20] Finally, détente proved of little help in ending America's longest war, Vietnam. That war finally closed in 1975 not because Beijing or Moscow pressed North Vietnam and the Vietcong to settle it, but because the latter triumphed against a weak, self-serving, South Vietnamese government that could never command a significant popular following, and against the United States which had made a tremendous investment in men and money but could find no way to win against a determined adversary (see Chapter 14).

Although détente initially won favor in the United States, the methods Nixon and Kissinger used to pursue their foreign policy stirred up so many political storms that the domestic base of support crumbled. Some critics protested the escalation of violence in Vietnam through accelerated bombing raids and the 1970 invasion of Cambodia; others charged that they had sold out to the Communists by making peace with them in 1973 and then letting Saigon fall two years later. SALT-I drew critics, too: People who believed in nuclear supremacy could not accept nuclear parity or any agreement that demonstrated some trust in the Soviets. Others lamented that SALT-I did not go far enough in controlling the arms race. Nor was everybody happy that in the 1970s the United States broke records for arms sales abroad ($10 billion in 1976 alone). Nixon and Kissinger also conducted much of their foreign policy in secret. Members of Congress were seldom invited into the inner circle of decision-makers. Congress passed the 1973 War Powers Resolution to restrict foreign adventurism, and it denied the White House funds for certain interventions. Legislative-executive relations became frosty, damaging détente. Kissinger, an Assistant to the President for National Security Affairs for 1969–75 and Secretary of State for 1973–77, preferred personal diplomacy; he would often side-step normal State Department channels, creating friction with resentful career diplomats. Critics believed too that Kissinger followed the ruthless maxim that the ends justify the means. He ordered wiretaps on the telephones of his aides and journalists because unauthorized leaks (Kissinger was a notorious leaker himself) were breaking his seal of secrecy. He sponsored CIA plots that held America up to ridicule for advocating democracy but actually undermining it. He defended President Nixon in the lowest days of the Watergate scandals. For Kissinger, Watergate and Nixon's ignoble resignation in 1974 became the simple reasons why he could not accomplish more. "Domestic divisions," he complained, had become a danger to his foreign policy, including détente.[21]

Détente achieved successes, but in the end promised more than it could deliver. Nixon and Kissinger over-estimated the importance of the triangular relationship of China, Russia, and America, and especially the usefulness of China as a check on Russia. Their grand strategy assumed wrongly that the Russians could manage their "friends" in North Vietnam or India or the Middle East and that Third World troubles derived from, and could be calmed by, great power decisions. The Nixon Administration still saw small states as proxies of the great powers and thus interpreted events largely through a Cold War perspective. In other words, the strategy paid too little attention to the local sources of disputes and the fierce independence of nationalist and neutralist governments. Kissinger spent much of his time trying to keep détente glued together against the backdrop of violent conflicts in Asia, Africa, and the Middle East, and economic challenges from OPEC. Even America's friends caused difficulty: Iran insisted on huge arms shipments but raised oil prices, threatening the American economy, and Saudi Arabia demanded sophisticated weaponry but refused to help resolve the explosive Arab-Israeli conflict. Above all else, détente failed to temper the American penchant for exaggerating the Soviet/Communist threat. Hence it failed to educate Americans about the necessity of measuring the threat exactly and meeting it soberly and patiently through diplomacy.

# 13

## The Clandestine Response:
## The CIA, Covert Actions,
## and Congressional Oversight

His temper flaring, Senator Daniel P. Moynihan resigned his Vice Chairmanship of the Select Committee on Intelligence, complaining that the Central Intelligence Agency had not properly briefed the committee as required by law. "I am pissed off," Senator Barry Goldwater lectured the CIA director at the same time. "This is an act violating international law. It is an act of war. For the life of me, I don't see how we are going to explain it."[1] In the spring of 1984 the CIA's mining of Nicaragua's harbors and the subsequent destruction of foreign merchant ships had stirred Congress. But Moynihan soon withdrew his resignation, and Goldwater cooled down after CIA Director William Casey offered profound apologies for having neglected Capitol Hill.

Yet clashes between intelligence officials and congressional overseers stormed again and again, as in May of 1985 when a car bombing in Beirut, Lebanon, killed more than eighty people. The massacre was carried out by agents working for a Lebanese group that was in turn working for the CIA. The new Vice Chairman of the Senate Intelligence Committee, Patrick J. Leahy, protested that the CIA's counter-terrorism program in the Middle East had not been adequately explained to him; he had discovered the covert operation on his own. Senator Moynihan allowed that he had been officially notified some months before about the secret program, but that he had not pushed for details. Many wondered not only if the CIA had once again lost control of foreigners whom the agency helped organize, fund, and supply with arms, but also had once again hoodwinked Congress.

Whether the question is the CIA's studied reluctance to inform Congress about covert actions or Congress' own laxity in pressing for information, congressional oversight of intelligence activities since the 1940s has been highly controversial, often flawed and ineffectual, seldom alert, frequently tardy, perhaps impossible, and yet certainly necessary. One newspaper cartoonist depicted the problem graphically when he sketched a sign in front of CIA headquarters; the acronym "CIA" was spelled out as "Congress Isn't Aware."[2] Representative Wyche Fowler of the House Permanent Select Committee on Intelligence has aptly remarked that "it is not oversight if it comes as an afterview."[3] Since the founding of the CIA in 1947, in fact, "afterview" has been the more common characteristic.

When he was criticized for first aiding and then abruptly abandoning the Kurds in Iraq, contributing to the deaths of thousands of them, Secretary of State Henry A. Kissinger coldly replied that "one must not confuse the intelligence business with missionary work."[4] Indeed, we are not dealing here with "missionary work," monotonous legislative history, or abstract constitutional questions. Covert actions is a murderous business. People of all ages and both sexes suffer and die as victims of CIA operations. Assassinations, bombings, torture, arson, paramilitary expeditions, and various "dirty tricks" maim, kill, and destroy—and they hurt the innocent. For example, in 1971 a covert operation introduced into Cuba, through a sealed vial, the African swine fever virus. The outbreak of swine fever a few weeks later decimated the hog population and caused severe pork shortages for the Cuban people (pork was a major component of the Cuban diet).

Some veteran public servants and analysts like George W. Ball and George F. Kennan have argued that covert action—as distinct from the collection of intelligence data—can never be kept secret, seldom attains policy objectives, undermines diplo-

macy, and is fundamentally incompatible with democratic principles and ethical values. Covert operations have exposed the
basic tension between open and closed government and between public accountability and secrecy. Public review, most
would agree, is essential if citizens are to acquire the knowledge
necessary to change policy and make leaders responsive. For all
of these reasons, some critics have called for strict limits on or
the outright abolition of covert actions. To study the problem of
congressional oversight, then, is to probe matters of life and
death and of great national consequence and to demonstrate the
extreme extent to which American leaders were willing to go to
meet the Communist threat they thought omnipresent.

The story has evolved through three stages. First, from 1947
to the early 1970s, when a series of shocks exposed the delin-
quency of congressional oversight. Second, from the early 1970s
to 1980, a period called by some the "oversight revolution,"
when Congress investigated past covert operations, found
much wrong-doing, created new watchdog committees on intelligence, and required the executive branch to keep Congress
better informed and more closely involved in deciding whether
covert actions should be undertaken.[5] And, third, the 1980s,
when the Reagan Administration and the new oversight process
coincided—and collided—to raise doubts about whether the
reform of the 1970s amounted to anything more than a temporary respite from congressional timidity and laxity on intelligence questions.

There are basically three kinds of intelligence activities: espi-
onage, counter-intelligence, and covert action. The first is the
gathering and analyzing of information, sometimes through
spying. Neither this activity nor counter-intelligence—the protection of information and the spy system—have been major
points of contention, although members of Congress sometimes
have complained that the CIA does not predict accurately from
the information it has collected. But, covert action has aroused
the most controversy. Covert action seeks through manipulation to influence governments, organizations, and individuals
abroad so that they support United States foreign policy. Such
secret activities have included not only propaganda, disinfor-
mation, bribery, and economic sabotage, but also paramilitary

operations wherein weapons and force are used. In short, covert action has often constituted the waging of war. On this the Constitution is explicit: *Congress* must declare war. It does not matter, as Senator Thomas Eagleton once explained, whether Americans who go to war wear green uniforms or seersucker suits. War is war. And "from little involvements—little CIA wars, big wars grow," he warned in an unsuccessful attempt to persuade Congress to extend the provisions of the 1973 War Powers Resolution to cover covert action.[6]

From the creation of the CIA in 1947 to the early 1970s, the first stage of this history, the agency expanded from its initially intended mission of information gathering to covert action. Although Congress had (and still does have) an obligation to monitor executive activities and to control the budget—specifically to insure that the CIA meets acceptable legal standards and is accountable for its actions, Congress seemed content to leave oversight responsibilities to the executive branch. But executive oversight proved woefully inadequate, despite the fact that the President, Special Assistant for National Security Affairs, National Security Council, 40 Committee, Department of State, American ambassadors, and President's Foreign Intelligence Advisory Board were well-positioned to oversee CIA activities and to check abuses. In some cases, Presidents themselves may not have known what the CIA was doing—including its plots to kill Cuba's Fidel Castro and the Congo's Patrice Lumumba. In the 1960s, Secretary of State Dean Rusk was sure that *he* knew; but when asked later if he had been given thorough information about covert actions, he answered: "I must confess at the time I thought the answer was yes, but it turned out not to be yes, and I feel very badly about that."[7] He recalled that the NSC, of which he was a member, did not even conduct an annual review of the CIA budget; indeed, he never saw a CIA budget.

"There are some things that you don't tell Congress; some things you don't even tell the President," admitted a CIA official.[8] Apparently CIA officials did not even tell the CIA Director himself about some projects. For example, counterintelligence chief James Angleton built a mini-empire within the CIA for twenty years until he was fired in 1974. One of his more abusive activities in the early 1960s was the three-and-a-half-year imprisonment of Soviet defector Yuri Nosenko. Angleton suspected Nosenko was a double agent sent to the United States to spy. For 1,277 days Nosenko was tormented in an austere, solitary prison near Washington, D.C. Apparently he was not a Soviet spy at all. Not until 1976 did an internal CIA study reveal the travesty. And when John McCone became Director in 1961, William Harvey, the CIA officer who was plotting with crime bosses to kill Castro, deliberately withheld information on the murder plots from McCone.

If the President and CIA Director were not privy to all of the agency's covert operations, it is hardly surprising that Congress knew even less. But Congress did not want to know. Senator Leverett Saltonstall expressed a widely-held view when he remarked that "it is not a question of reluctance on the part of CIA officials to speak to us. Instead it is a question of our reluctance, if you will, to seek information and knowledge on subjects which I personally . . . would rather not have."[9] Senator Goldwater added years later that "we would be better serving our country by not hearing it."[10]

Why did most members of Congress accept "deliberate ignorance?"[11] Why did they prefer not to know? In the heady days of the Cold War mentality and "imperial presidency," the CIA won favor as a presidential instrument for meeting the Communist threat—and it should not be fettered. Members of Congress did not want to appear to impede the agency's anti-Communist activities by raising potentially embarrassing questions. In some cases, they shied away from moral issues raised by some of the CIA's dirtiest tricks. Ignorance may not always be bliss, but it can provide escape. Representatives and senators also preferred ignorance because they worried about breaching a national security secret through an inadvertent public utterance. Under the lights and in front of the microphones, some legislators found it difficult to recall which information was classified and which was not. Others resisted secret information because they feared co-optation. If they received notice of a covert action and did nothing, this could be interpreted as acceptance. And if the covert action later became controversial, they would be hard pressed to explain why they had earlier failed to prevent it. This point is closely related to another: there was little political payback from activist congressional oversight. The information was confidential, hearings or briefings were held behind closed doors, and legislators were not supposed to go public.

The Congress's hesitancy to delve into CIA operations was conspicuously evident in the system of congressional oversight. With no legal basis, this system was highly personalized and episodic, responsive to the whims of the CIA Director and a handful of leaders in the two houses. Someone tagged the system with the acronym BOGSAT—"bunch of guys sitting around a table."[12] Information on the CIA and its budget, and hence oversight, was restricted largely to subcommittees of the Appropriations and Armed Services Committees of both houses. Powerful, senior senators like Richard Russell, John Stennis, and John McClellan dominated the structure. One intelligence official noted in an internal CIA report that "they tend generally toward conservatism and hawkishness."[13] CIA Director William Colby appreciated them, for they "faithfully and patriotically" protected the CIA from "public prying."[14] In Russell's words, the CIA's methods had to be taken "on faith."[15]

This senior command with oversight responsibilities occasionally received CIA briefings, but usually only after covert action had been initiated. The subcommittees lacked adequate staff and seldom convened. Attendance was poor, and members asked few questions of intelligence officials. Nor did the subcommittees issue reports. "We met annually—one time a year, for a period of two hours in which we accomplished virtually nothing," observed Representative Walter Norblad, a House Armed Services intelligence subcommittee member for four years in the 1960s.[16] In some years the Senate Armed Services intelligence subcommittee did not meet at all. One CIA official who attended sessions reported to his superiors that the members were "prone to intermittent dozing" and "failing faculties." He mockingly told the story of the elderly chairman who, when shown a diagram of covert actions, demanded to know "what the hell are you doing in covert parliamentary operations." When reassured that the chart read "paramilitary operations," he shot back: "The more of these the better—just don't go fooling around with parliamentary stuff—you don't know enough about it."[17]

Moreover, these subcommittees became cozy with the CIA, in part because they prided themselves on being its protectors but also because the CIA cleverly cultivated them. The problem, of course, is a perennial one: regulators often come to reflect the views of the institution they are supposed to be regulating. "The clandestine services give them a peek under the rug and their eyes pop," a CIA official remarked. "It doesn't take long before the Congressional overseers acquire that old-school feeling."[18] Sometimes subcommittee chairmen became "insiders" when they visited CIA headquarters for special breakfasts. In the 1950s Director Allen Dulles fed tidbits of information about secret projects to the select few. "Allen used to find the congressmen were intrigued with little personality stories and quasi-clandestine details which would amuse [them]. But I think he found it an effective way of building a rapport with them," a former CIA officer commented.[19] But, as Representative Michael J. Harrington later complained, members of Congress became "accomplices," because "the more they know, the more they are responsible for hiding."[20]

If the Congress was both ill-informed and lax about oversight, the CIA was determined to keep it that way. Agency officials followed the practice of never volunteering information, and if asked about an operation, to provide a minimum of details. In some cases, the CIA ignored Congress altogether, flatly refused to answer questions, or told lies in briefings. CIA officer Ralph W. McGehee has written that in 1964 he helped prepare a briefing on operations in Laos that Far Eastern Division Chief William E. Colby presented in secret to Congress. According to McGehee, "it was a complete hoax contrived to decieve Congress, which naturally swallowed it hook, line, and sinker."[21] Allen Dulles once told a colleague: "I'll fudge the truth to the oversight committee, but I'll tell the chairman the truth—that is, if he wants to know."[22] CIA officials feared leaks, but more, they believed that Congress had no right to information. When Senator J. William Fulbright, chairman of the Foreign Relations

Committee, asked the CIA if the Fulbright academic awards were ever used as covers for CIA activities, the agency would not tell him.

An instance demonstrating well the combination of Congress's reluctance to know and the CIA's reluctance to tell occurred in 1971 in the Senate. The issue was the CIA's 1960s covert war in Laos—a $5 billion program using over 10,000 Meo tribesmen as mercenary troops. Congress never directly voted the funds—money from the Defense Department and Agency for International Development had been secretly transferred to the CIA—but apparently some senators had been briefed. When information about the covert war surfaced, Senators Fulbright and Alan Cranston confronted Senator Allen Ellender, an overseer:

FULBRIGHT  Would the Senator [Ellender] say that before the creation of the army in Laos they [CIA] came before the committee [intelligence subcommittee of Appropriations Committee] and the committee knew of it and approved it?

ELLENDER  Probably so.

FULBRIGHT  Did the Senator approve it?

ELLENDER  It was not—I did not say anything about it. . . . It never dawned on me to ask about it. I did see it publicized in the newspaper some time ago.

CRANSTON  The chairman stated that he never would have thought of even asking about CIA funds being used to conduct the war in Laos . . . I would like to ask the Senator if, since then, he has inquired and now knows whether that is being done?

ELLENDER  I have not inquired.

CRANSTON  You do not know, in fact?

ELLENDER  No.

CRANSTON  As you are one of the five men privy to this information, in fact you are the number one man of the five men who would know what happened to this money. The fact is, not even the five men know the facts in the situation.

ELLENDER  Probably not.[23]

Another senator, Stuart Symington was outraged to learn about the Laos operation from the press: "The large majority of

Congress was deceived by Laos. . . . We now get reports that Americans died in Laos and that we had been lied to, because we were told they died in Vietnam. We are getting pretty sick of being lied to. . . ."[24]

Unlike that for any other government agency, the budget for the CIA was kept secret from Congress—except from the few overseers. Although the Constitution prescribes that "a regular statement and account of the receipts and expenditures of all public money shall be published" (Article I, Section 9), the CIA enjoyed covert financing. The CIA budget was concealed in the figures for other departments—usually the Defense Department—and then moneys were transferred to the agency. Congressman Ed Koch once asked Director Richard Helms about the size of the CIA budget. Helms said the CIA did not answer such questions, members of Congress had no right to know, and the CIA budget was properly hidden. "Do you mean that it might be included under Social Security?" asked the irritated representative. Helms's and the CIA's arrogance then shone: "We have not used that one yet, but that is not a bad idea."[25] Moreover, the agency's "Reserve for Contingencies," created in 1952, permitted the CIA great flexibility in using its dollars for covert operations without congressional scrutiny. In short, there was almost no accountability. Lucien Nedzi, a House subcommittee member, told his colleagues: "I have to be candid and tell you I don't know whether we are getting our money's worth."[26] Congress has always voted down bills to disclose intelligence budget figures.

This state of affairs—shallow congressional oversight and CIA secrecy and deceit—did not go unchallenged. Indeed, in the period 1947–74 about one hundred and fifty proposals to improve oversight were introduced in Congress—many called for the creation of joint or select intelligence committees—but only two bills ever reached the floor or Congress. In the mid-1950s Senator Mike Mansfield asked for closer scrutiny of an agency he considered uncontrolled. In 1956 his reform proposal earned the bipartisan support of thirty-four co-sponsors, and the Rules Committee passed it by an 8 to 1 tally. Besides setting up a joint congressional committee with a professional staff, the resolution required the CIA to keep the committee fully informed. The chief argument for the new structure was that a specialized committee would permit Congress to concentrate on intelligence with the same degree of professionalism and thoroughness it gave to other subjects and thus meet its obligations under the checks-and-balances system. From such congressional surveillance, it was also believed, a better intelligence service would emerge. But freshman Senator Mansfield collided head-on with the "high brass" of the upper chamber.[27] They defended as workable the system they themselves dominated; they countered that intelligence should not be "watchdogged to death."[28] The CIA itself fought the legislation. As Mansfield later recalled, CIA officials were able to defeat reform bills because "they had the hierarchs in the Senate in their pockets. . . ."[29] Losing several of its co-sponsors, the Mansfield resolution suffered defeat, 59-27. Four years later, at a White House meeting with President Dwight D. Eisenhower, Mansfield again advocated a congressional oversight committee. Senator Russell complained that the committee's staff might leak secret information that would endanger the lives of intelligence officers, and Mansfield's suggestion was quickly "knocked down" by other senators present.[30]

Not until 1966 did another oversight resolution reach the Senate floor, although many had been introduced. Then, Senator Eugene McCarthy sought to end what he called the "invisible government" of the CIA that was reponsible for the Bay of Pigs and the Vietnam War.[31] McCarthy proposed that some members of the Foreign Relations Committee be invited into the oversight process. The chairman of that committee, J. William Fulbright, backed the McCarthy initiative, but Senator Russell, chairman of Armed Services, once again blocked reform. In a testy debate with Fulbright, the Georgia Senator warned him against "muscling in on my committee."[32] Russell persuaded the Senate to refer the resolution to his committee, where he killed it. The intent of the motion was nonetheless satisfied in January 1967 when Russell invited Fulbright and two other members of Foreign Relations to attend meetings of the Armed Services intelligence subcommittee. But this new system served oversight only minimally: still a mere handful of senior legislators held responsibility; still legislators heard largely what

the CIA wanted them to hear; still covert actions went unques-
tioned and unknown; and still the CIA budget remained secret.
In other words, Congress could not determine if the CIA was
meeting its statutory mandate—or even meeting the Communist threat.

The feebleness of oversight continued until the early 1970s
when the Vietnam War, Watergate scandals, and revelations of
unsavory CIA covert actions at home and abroad weakened the
executive branch's ability to resist a stronger congressional voice
in foreign policy and intelligence. Congressional oversight is
alert and active, it appears, only when a foreign policy consensus is shattered, when executive authority is vulnerable, or
when investigatory journalism has stimulated uncommon public concern.

The origins of the so-called "oversight revolution" owed
much to Watergate, that array of scandals that ultimately drove
Richard M. Nixon from office in shame. Watergate induced a
temporary shift of power from the executive to the legislative
branch. The Vietnam War after the 1968 Tet Offensive, various
reports published in the early 1970s on the CIA's role in the
removal of the Diem government and the secret war in Laos,
and the passage in 1973 of the War Powers Resolution—all had
already begun this rearrangement of governmental power.
Some of the burglers arrested for the 1972 break-in at Democratic National Committee headquarters were former CIA personnel, and the CIA had participated in the Watergate cover-up
by issuing cover stories designed to impede the FBI investigation. The CIA had also helped Nixon politicos in various illegal
activities at home. Several committee reports in 1973–74 chided
the CIA for its transgressions, including the destruction of
records related to Watergate.

Events in Chile accelerated oversight reform. In October 1973,
President Salvador Allende was overthrown in a military coup.
Rumors quickly circulated that the CIA had helped undermine
the elected Marxist leader. Under oath, American officials,
including Secretary Kissinger, assured Congress that the allega-
tions were false. But, then, in April 1974, as suspicions were
taking on more substance, CIA Director William Colby told a
closed session of the intelligence subcommittee of the House

Armed Services Committee the truth about the CIA's considerable covert program to topple Allende. Representative Harrington, who considered oversight a "sham," urged House leaders to conduct hearings on the Chilean affair.[33] Getting nowhere, in part because Congress was so absorbed with the process of impeaching President Nixon, but having obtained Colby's report from the subcommittee chairman, Harrington decided to break the rules and speak out. "I couldn't believe my eyes. . . . There it was, 40 pages in black and white . . . telling in clinical detail how we were engaged up to our eyebrows [in Chile]."[34] At a September 12, 1972, press conference the congressman from Massachusetts revealed Colby's testimony. Embarassed veteran overseers like Senator John Stennis admitted that they had not known much about the CIA's role in Chile. In fact, of the thirty-three CIA covert projects in Chile during 1963–73, only eight had been reported to the appropriate congressional committees. "I do not think there is a man in the legislative part of the Government who really knows what is going on in the intelligence community," Senator Howard Baker remarked. "I am afraid of this lack of knowledge. For the first time, I suppose, in my senatorial career I am frightened."[35]

Apprehensions like Baker's helped pass the Hughes-Ryan Amendment to the 1974 Foreign Assistance Act, the first significant congressional oversight reform. Hughes-Ryan held that the CIA could engage in a covert action abroad only if "the President finds that each operation is important to the national security of the United States and reports, in a timely fashion, a description and scope of such operation to the appropriate committees of the Congress, including the Committee on Foreign Relations of the United States Senate and the Committee on Foreign Affairs of the House of Representatives."[36] The Hughes-Ryan reform thus required reports on covert action to Congress; and it expanded the number of overseers—now to include members of six committees. The measure did harbor shortcomings: The CIA did not have to report to oversight bodies *before* implementing an operation; the reports were not required to be thorough; and after receiving classified information in the vaguely mentioned "timely fashion," congressional overseers still had no veto power unless they were bold enough to violate the pledge of secrecy and halt CIA covert action by denying funds through a vote of the entire Congress. Nonetheless, this new reporting procedure apparently had a "chilling effect" on clandestine actions, which dropped dramatically in number.[37]

The Hughes-Ryan Amendment hardly quieted the intelligence tempest, for in late December 1974 the *New York Times* revealed extensive CIA spying on Americans at home in violation of the agency's statutory charter. Soon James Angleton and other CIA influentials were fired; and President Gerald Ford appointed a commission to investigate the charges. What would Congress do this time? On January 27, 1975, by an 82 to 4 margin, the Senate created a new committee—the Select Committee to Study Governmental Operations with Respect to Intelligence Activities. Soon known as the Church Committee after its chairman Frank Church, this investigative body hired a professional staff, established guidelines to maintain the security of classified materials, held extensive hearings, and produced a shelf of publications detailing CIA activities. In July, the House also created an investigative committee chaired by Otis Pike.

While the two committees probed and pressed, shocking public opinion leaders with information on assassination plots and illegal mail openings, the Hughes-Ryan Amendment encountered its first major test. The issue: CIA covert action in Angola, that African nation just about to be liberated from colonial bondage by Portugal and already wracked by civil war. Before 1975 the United States had backed one faction and the Soviets another. In January of that year the Ford Administration decided to increase the stakes through more covert political aid. The following month the leaders of the six congressional committees received notification; they apparently made no fuss. Then in July the administration opted for covert military assistance to two Angolan factions. Senior members of the congressional committees duly received CIA notices. This time, however, Senator Dick Clark of the African Affairs Subcommittee of the Foreign Relations Committee asked for more details, which were given to his group in a briefing. Controversy surrounds this briefing and others, for at least two former CIA officers have admitted that they were neither complete nor accurate. Clark, especially after an August trip to Africa, grew worried that the United States was rejecting a viable diplomatic alternative and exaggerating the Soviet threat in Angola. About the same time, the Assistant Secretary of State for African Affairs, Nathaniel Davis, quietly resigned with much the same reasoning after his futile attempt to stop covert military aid.

Still the President wanted more; in September he notified the appropriate committees that military aid was being hoisted to the level of twenty-five million dollars. Clark began to lobby the Foreign Relations Committee, which was becoming wary of deeper American involvement in volatile Angola. When, in December, the administration informed the committees that an additional $7 million was being sent as covert military aid to Angola, the senator from Iowa decided to go public. He thereupon persuaded his Foreign Relations colleagues to report an amendment to the 1976 foreign assistance bill to prohibit any funds for covert action in Angola. Legislative jockeying and political opportunism converted the Clark Amendment into the Tunney Amendment to the Defense Appropriations bill, where, it was assumed, the CIA's budget for its Angolan operation rested. On December 19, the Senate, by a count of 54 to 22, barred covert action in Angola; the House agreed on January 27, 1976 (323 to 99); and the President reluctantly signed the act on February 9. An obstinate Congress had halted a covert action and had as well exposed the weaknesses of Hughes-Ryan oversight. Some legislators began to ask for more reform—for notification before projects were launched and greater control over covert actions.

But the case for stronger congressional oversight sputtered in February 1976 because of the way the final report of the Pike Committee was handled. Pike's committee completed its work in January, yet the full House refused to publish its report until the administration had a chance to give it a classification review. Pike thereupon released a brief summary of the committee's recommendations only; but a draft of the final report came into the hands of the CBS-TV reporter Daniel Schorr, who passed it on to the *Village Voice*. The publication of excerpts of the report in this weekly New York newspaper distracted attention from the committee's findings on CIA abuses and inadequacies and put it on the source of the leak (which was never determined) and Congress's leakiness. The Pike Committee urged, among other changes, the creation of a new committee on intelligence to which the CIA Director would report fully on any covert action within forty-eight hours of the decision to initiate the action.

The impressive final report of the Church Committee, published in April 1976, encouraged oversight reformers. After detailing excessive presidential use of covert action, the Church Committee asked the Senate to increase the congressional role in intelligence through a legislative charter; to publish an overall intelligence budget figure; and to create a new Senate committee on intelligence. The Senate soon voted, 77 to 22, to create the Select Committee on Intelligence. (The House waited until July of 1977 to launch a similar committee.) Because Hughes-Ryan was not repealed, the CIA now had to report to eight committees. But the new system faced an uncertain future: vocal opponents began a counter-attack, the question of a charter remained unresolved, and the incoming Jimmy Carter Administration had its own ideas.

President Carter moved quickly to reorganize the maligned CIA, reduce its bloated bureaucracy, improve its intelligence production, and restrict its activities. His administration was not adverse to covert action; Carter officials, CIA Director Stansfield Turner recalled, "turned easily and quickly to covert devices . . ." especially in its last two years.[38] On the whole, congressional intelligence committees were properly notified, to the effect that some planned actions were scaled down and at least one stopped. In a 263-page bill titled the National Intelligence Reorganization and Reform Act of 1978, oversight advocates offered an elaborate statutory charter for the intelligence community, including prohibitions on assassination. Critics pilloried the legislation as both too restrictive and too permissive; and the administration cooled toward it. Then the taking of the American hostages in Iran and the Soviet bludgeoning of Afghanistan further undermined the oversight reform movement. Why cripple an already weakened CIA when it was urgently needed as a weapon in the post-détente Cold War struggle with the Soviet Union? critics asked.

New but shortened charter legislation was introduced in early 1980. The first administration witness, Director Turner, nonetheless singled out several provisions for complaint. The bill required the CIA to notify congressional intelligence committees before starting a covert action. Turner found this a detriment to presidential flexibility at times when speed and secrecy were paramount. In general, the Carter Administration believed that a statutory reporting requirement was an "excessive intrusion by the Congress into the president's exercise of his powers under the Constitution."[39] The basic issue was joined once again: The President vs. the Congress in matters of foreign policy.

Several weeks after this testimony, the kind of problem Turner had identified arose in the covert expedition to rescue the hostages in Iran. President Carter worried about maintaining secrecy for the complicated mission, so he planned to notify congressional leaders only after "the rescue operation had reached the point of no return." But because the mission had to be aborted while in progress, the President "never got around to that."[40] Given the emergency nature of the problem, congressional overseers gave the President the benefit of the doubt, but continued to argue for prior notification.

With Iran as a backdrop and with critics such as Republican presidential candidate Ronald Reagan charging that the CIA was being hamstrung by burdensome restrictions, Congress jilted charter legislation in favor of a two-page document—the Intelligence Oversight Act of 1980. The new legislation required the CIA to report only to the House and Senate Intelligence Committees (thus revising Hughes-Ryan) and to give prior notification *except* when the President determined that "extraordinary circumstances affecting vital interests" existed, in which case only eight top leaders of Congress would be notified. And the President could actually *waive* prior notification altogether if later, "in a timely fashion," he explained his reasons. Finally, the act required the CIA to furnish any information—classified materials included—requested by the two committees, and to report "any illegal intelligence activity or significant intelligence failure. . . . "[41] This legislation hardly satisfied those who preferred a major statutory charter; it carried a certain ambiguity permitting the President to skirt prior notification; it placed no time limit on covert actions and did not provide for a congressional veto (in this regard it was different from the War Powers Resolution). In essence, then, Congress only had to be informed.

The new Reagan Administration greatly expanded CIA covert actions. At the same time, it publicly undertook covert operations that in earlier periods probably would have been kept secret: military aid to Nicaraguan *contras*, Cambodian rebels, Afghan insurgents, and Angolan forces—all "freedom fighters," said the President; "terrorists" replied his critics. In part this practice stemmed from Washington's eagerness, under the Reagan Doctrine of aiding anti-Communist guerrilla groups, to send an unmistakable signal to the Soviets and their clients and allies that the United States would contest them everywhere. But it also derived from an awareness that secret wars seldom remained secret. In this atmosphere, the Reagan Administration showed minimal respect for congressional oversight and the briefing process. As well, it preferred executive orders to a statutory charter. But, more, Congress in the 1980s retreated from the reform mood of the 1970s, coming to focus "on enhancing effectiveness" rather than "on circumscribing [the CIA's] capabilities and activities."[42] In 1985, for example, Congress repealed the 1976 Clark (or Tunney) Amendment banning covert action in Angola. The following year the Reagan Administration informed Congress that it was extending $15 million to a rebel group led by Jonas Savimbi to help it overthrow the Angolan government. Talks between Angola and the United States to resolve the issue of Namibia soon broke down.

Still, some legislators protested Reagan's approach to covert action and oversight. One Democratic member of Congress complained that, to the reinvigorated CIA, "we are like mushrooms" because "they keep us in the dark and feed us a lot of manure."[43] Another lamented that Congress had "little leverage" over the intelligence community. "You don't have a veto, so you have to satisfy yourself by hollering inside the tin can."[44]

A critic described CIA briefings of intelligence committees as a "tooth-pulling process, and sometimes the dentist can't see all the teeth."[45] According to a Republican member of Congress, CIA Director William J. Casey "wouldn't tell you if your coat was on fire—unless you asked him."[46] Indeed, Casey frankly lectured congressional overseers: "I do not volunteer information. If you ask me the right question I will respond."[47]

In the Reagan eighties, Congress found it difficult to halt a covert action it disliked. When notified by the President and CIA of an impending covert activity, the intelligence committee members could ask questions, press for changes, demand termination, or contact the President and register doubts with the White House. If such pleading failed, a member could attempt to amend a money authorization bill to prevent the expediture of funds for a particular clandestine project. In short, a member of Congress had to be willing to confront or embarrass a popular President with formidable rhetorical and patronage powers. It was not a task many took on.

Congress supported most of Reagan's covert operations, but Nicaragua proved particularly nettlesome. A Marxist government struggled to launch its revolution at home as the CIA funded and trained insurgents (the *contras*) in Honduras to topple the new regime that had welcomed Soviet and Cuban assistance as a counter to the hostile United States. In September 1982 the Boland Amendment banned covert military aid to the *contras* "for the purpose of overthrowing the Government of Nicaragua."[48] The administration partially side-stepped the restriction by stating that the aid was intended to interdict arms sent by Nicaragua to anti-government insurgents in El Savador. The *contra* program, moreover, was constantly enlarged, including the mining of Nicaraguan harbors. The Senate passed a non-binding resolution in April 1984 (84-12 vote) asking that no funds be expended for the mining of the ports; the House did the same (281-111). Congress also forbade any intelligence agency from conducting military or paramilitary operations against Nicaragua as of November 8, 1984. But in the summer of 1985 Congress acceded to presidential pressure and approved "humanitarian" aid to the *contras*. That summer, too, the administration admitted that an official of the National Security

Council was giving or had given direct military advice to the
*contras.* "We're not violating any laws," claimed President
Reagan.[49] But Representative George E. Brown, Jr., a member of
the House Intelligence Committee, complained that the CIA has
"been giving lip service to the Boland Amendment. The [admin-
istration's] legal eagles are interpreting all of the laws in a way
to favor the policies of the President."[50] Then, in the summer of
1986, Congress approved $100 million for the *contras,* providing
highly publicized aid to a supposedly covert operation against a
country, Reagan charged, that had gone Communist.

Perhaps congressional oversight can seldom be satisfactory or
effective. Obstacles remain formidable. First, oversight seems to
be most ardent only when the press and then the public opinion
elite are stimulated to debate foreign adventurism. But the
extreme secrecy that shrouds CIA activities, the intelligence
budget, and the oversight process itself prevents such moments
from occurring very often and stymies the public airing so vital
to governmental accountability. In 1984, Congress actually
succumbed to the CIA's preference for more secrecy when it
passed the Central Intelligence Agency Information Act, which
exempted operational files from the Freedom of Information
Act. Second, presidential power dominates most foreign policy
debates, and precedent enhances presidential initiative and
prerogative in times of crisis. To put it bluntly, the executive
branch can get away with murder: It can quietly expand a
program beyond what members of Congress originally intended, or it can redefine a problem to circumvent congressional
oversight. In 1981, for example, the CIA secretly sent a team into
Laos in a vain effort to find Missing-in-Action soldiers from the
Vietnam era. Director Casey did not give Congress prior notification because, he claimed, the mission was not a covert action
but an operation to gather information. Third, the great number
of intelligence bodies makes it difficult for Congress to monitor
the intelligence community and permits the executive branch to
utilize agencies that receive less congressional scrutiny. Fourth,
Supreme Court decisions have shielded the CIA from public
scrutiny.

Fifth, the persistence of the Cold War and virulent anti-
Communism, replete with exaggerated rhetoric about the Com-

munist threat, cascading crises, and impassioned appeals to patriotism favors giving the CIA a wide range within which to roam. When détente is down and confrontation is up Congress tends to retreat from careful oversight and to accept "dirty tricks" as a necessary evil. Sixth, vagueness is a characteristic of much legislation, including the Intelligence Oversight Act of 1980. It is not clear when prior notification must be given or when it can be waived. This ambiguity affords the executive branch considerable latitude. Seventh, some members of Congress prefer not to know about the CIA's covert business, for the reasons discussed earlier. Eighth, members of Congress are "busy generalists," too overwhelmed by multitudinous assignments fully to watch over the CIA.[51] Ninth, given the restraints imposed by secrecy, access to knowledge about the CIA budget is obstructed. The House Permanent Select Committee on Intelligence, for example, has made available for inspection to members of Congress a classified report on CIA activities at budget time. But very few representatives ever perused the looseleaf notebook. "Nobody reads the stuff, or almost nobody," observed one. "The hard thing is that this material is not accessible to your staff, so you don't have them review it," said another.[52] Because they are so busy, members of Congress in essence cast blind votes. And, finally, there is little political reward from controversial vigilance over intelligence when compared to farm, health, or other issues. "I have to go now," Senator Hubert H. Humphrey abruptly announced during a hearing on covert operations in Chile. "I am trying to get jobs for 400 people in Minnesota today. That is a great deal more important to me right now than Chile."[53] The lesson from Senator Frank Church's experience is conspicuous; in 1980, just a short time after his highly publicized stewardship of the investigating committee, he was defeated in Idaho for re-election to a fifth term, in part because right-wing groups charged that his disclosures had seriously wounded the CIA and the war against Communism.

In the 1980s the congressional monitoring system created during oversight reform in the last decade did not follow the scenario reformers had written for it, and the CIA seemed quite "willing to go off on its own and risk getting caught by

Congress."[54] The Reagan Administration, of course, deliberately inhibited the process. But, also significant, the oversight system that reformers had hoped would restrain covert actions had actually helped to augment them. Legislators received earlier than ever before in the history of oversight more information about more covert operations from more agencies; legislators knew more about the CIA budget than ever before. The CIA generally answered to the Congress. The Congress on the whole liked the answers. Indeed, members of Congress were using the information not to question or thwart secret missions but to support them, not to probe ends but to query means. Like predecessors in the 1940s, 1950s, and 1960s, Congress in the 1980s favored an aggressive CIA to wage counterinsurgency and to meet the Soviet threat in the revived Cold War. To the reformers' chagrin, then, oversight became an instrument for expanded covert action. "The CIA got what it wanted," Senator Moynihan said of his eight-year service on the Senate Intelligence Committee. "Like other legislative committees, ours came to be an advocate for the agency it was overseeing."[55]

Although critics of military and paramilitary covert actions continued to urge Congress to push beyond flawed oversight to outright abolition, a new crisis in the winter of 1986–87 showed that Congress would not abandon covert wars even though it was embarrassed that its oversight role was, once again, proven demonstrably feeble. The "Irangate" or "contragate" crisis revealed how contemptuously the Reagan Administration regarded the oversight process, how determined Reaganites were to overthrow the Communist threat they identified in Nicaragua, and how negligent congressional intelligence committees had been in monitoring clandestine activities. In 1983–85, despite repeated stories in the press, Congress did not investigate links between the National Security Council's Lieutenant Colonel Oliver North and private, right-wing groups who were sending arms to the *contras* after Congress prohibited the administration from aiding these anti-Sandinista forces. In 1986, the news hit the headlines: NSC and Colonel North had masterminded a covert sale of arms to a nation that applauded terrorism—Iran. Apparently the NSC staffers intended to gain the release of Americans held hostage by Islamic extremists in Lebanon, but only a few were set free. But the bungled covert deal did not stop there: profits from the arms sale were deposited in a Swiss bank account and then funneled to the *contras*, in clear violation of the law. President Reagan's popularity dropped quickly. He had approved the sale of weapons to Iran, a country most Americans considered an enemy, and he had ordered that Congress not be informed of the secret transaction—a clear violation of the Intelligence Oversight Act of 1980. If the President did not know that Colonel North and the NSC were running guns to the *contras*, then he appeared incompetent and out-of-touch with a major undertaking engineered in the White House itself. And if Reagan was privy to the diversion of funds to the *contras*, he knowingly violated Congress' ban on military aid to those forces. To many observers, familiar with Cold War history and the perennial over-reading of and over-reaction to "Communism," this episode stood as yet another example of the dangers posed by threat exaggerations to American domestic institutions.

# 14

## Ronald Reagan, Central America, and the Legacy of Vietnam

Soviet leaders, masters of an "evil empire," asserted President Ronald Reagan, stood prepared "to commit any crime, to lie, to cheat" in order to achieve a Communist world.[1] In a display of raw anti-Communism reminiscent of early Cold War days, Reagan rejected détente and claimed that an expansionist Soviet Union "underlies all the unrest that is going on. If they weren't engaged in this game of dominoes, there wouldn't be any hot spots in the world."[2] Reagan and his advisers identified Soviet intrigue as the source of Third World disorders. Quoting the Truman Doctrine and resuscitating the domino theory, they embraced familiar Cold War themes of bipolarism, global containment, and confrontation, altogether happy to discard the "Vietnam syndrome" and the limits it seemed to place on the exertion of American power. The Reagan Doctrine committed the United States to the active support of anti-Communist movements around the world, including the Nicaraguan *contras* and the Afghan rebels. Reagan determined to meet the Communist threat, which he believed omnipresent, head on.

The Reagan Administration's attention fixed on Central America, an area that had long suffered United States' influence and its own profound economic, social, and political divisions. Blaming turmoil in the region on the "Moscow-Havana axis,"

Reagan officials set out to defeat leftist insurgents in El Savador, topple the Sandinista government in Nicaragua, and draw Guatemala and Honduras into closer military alignment with the United States.[3] In El Salvador, the Reagan Administration said it had discovered "a textbook case of indirect armed aggression by Communist powers," and in Nicaragua it found a betrayed revolution—a country on the Soviet Union's "hit list," the Reaganites charged—that had gone "Communist."[4]

As Reagan predicted victory over "Communism" in Central America, critics predicted another Vietnam. "El Salvador is Spanish for Vietnam" read one bumper strip. The veteran diplomat George Ball, an in-house dissenter on Vietnam in the 1960s, dissented again; Reagan policy toward El Salvador, he insisted, was a case of "plagiarization."[5] Senator Christopher Dodd of Connecticut decried the "ignorance" of the Reagan officials—they seemed "to know as little about Central America in 1983 as we knew about Indochina in 1963."[6] Economic underdevelopment, poor medical care, illiteracy, poverty, and rigid class structure, not Soviet plotting, underlay Central American unrest, Dodd insisted. Reagan not only plunged the United States into civil wars and regional disputes in Central America and sought military solutions to social, economic, and political problems; he also rewrote the history of the Vietnam War, declaring it a "noble cause."[7] This time, in Central America, Reaganites bragged, the United States would stay the course and win, having drawn useful lessons from Vietnam—including a hesitancy to commit combat troops. Policymakers and critics alike, then, summoned the Vietnam legacy in their debate over United States intervention in Central America. Surveys of American leaders in 1976 and again in 1980 have revealed that their position on the Vietnam War provided the best predictors of their foreign policy beliefs. "Still in Saigon, still in Saigon, still in Saigon, in my mind," went the words of a 1981 song.[8]

But which Vietnam stayed in American minds? That is, which lessons did Americans learn from the war that might guide them into or out of Central America? No consensus has taken shape. At least four perspectives had become prominent by the early 1980s. First, the non-interventionist "no more Vietnams" school of thought: Never again should the United States intervene in a

Third World country undergoing national rebellion or civil war, because the United States cannot provide answers for problems indigneous to other peoples. Americans may have the money, but they do not have the knowledge, sensitivity, patience, or muscle to rearrange other governments, especially those distant from United States power. Let others learn from their own mistakes. Wisdom, then, lies in ending America's penchant for being the world's teacher, social worker, banker, and police-man.

A second perspective resembled the first but was more radical—the "inevitability" point of view. The United States will continue to suffer Vietnams, went this argument, so long as Americans must import scarce commodities like oil and minerals and export American products, and must therefore insure economic lifelines; so long as Americans persisted on remaining "Number One," whatever that meant; so long as Americans persisted in interpreting local crises derived from internal sources as Cold War confrontations demanding impositions of the containment doctrine; so long as Americans exaggerated the Soviet threat; so long as the imperial presidency distorts and overwhelms the checks-and-balances system.

The third perspective we might call the "diplomatic interven-tion" school: Because the United States is a great power with global interests that must be protected against all threats, and because Americans have an obligation to help others establish democratic, prosperous states, then the United States must be interventionist. But the means should be diplomatic rather than military. Thus, negotiate with leftist and nationalist regimes to reduce potential threats and use foreign aid to foster conditions compatible with American interests and principles. In this way, Americans can build a non-Communist world through the nurturing of nationalism—and guarantee American prosperity at the same time. President Jimmy Carter's initial openness to the leftist government of Nicaragua and his negotiation of the Panama Canal treaties seemed to reflect this thinking. Ameri-cans, said Carter, had to put their "inordinate fear of Commu-nism" behind them.[9]

The fourth perspective, often called the "win" school, emerged from American conservatives and military officers who served in Vietnam, and other Americans who came to believe that the Vietnam War could have been won. Some of them promoted a "stab in the back" thesis: Biased journalists, anti-war critics, and meddlesome members of Congress broke America's staying power and thus assisted the enemy. America supposedly lost its will. Civilian leaders hampered the military through gradualist methods rather than all-our war. "It takes the full strength of a tiger to kill a rabbit," lectured former General William C. Westmoreland.[10] "Remember," advised a former battalion commander, "we're watchdogs you unchain to eat up the burglar. Don't ask us to be mayors or sociologists worrying about hearts and minds. Let us eat up the burglar our own way and then put us back on the chain."[11] Too often, but especially after the Tet Offensive of 1968, these "win" advocates argued, the United States failed to follow up on military advantages to hurl defeat at the enemy and did not take the necessary step of invading North Vietnam. The military, Ronald Reagan concluded, was "denied permission to win."[12] In short, went this lesson, military answers are appropriate—and they can work. Others in this school of thought have argued that the war *should* have been won because it was a moral war—that is, it was quite moral for the United States to resist the tyranny of North Vietnam, whose aggression against Kampuchea and repression of Vietnamese after 1975 have demonstrated the Communists' brutalities.

Because Reagan foreign policy seemed to be fueled at least in part by the notion that Vietnam was fought in the wrong way and could and should have been won, and that Central America was not a potential Vietnam, the question of victory in the Vietnam War became a compelling issue in the 1980s. Could the war in Vietnam have been won? Would another 500,000 soldiers, billions of dollars more, unrestrained B-52 bombing raids, better counter-insurgency methods, additional equipment and perhaps even nuclear weapons—would all of this have delivered victory? Or would this only have meant that the United States lost more, destroying even more of what it was trying to save? Asked still another way: How could the Communist Vietnamese (the Vietcong in the South and Vietnamese of the North), outnumbered seven to one in military forces, lacking advanced weapons and bombers, using almost primitive methods of communication and transportation, and receiving little foreign aid (Russia gave only $1.8 billion, China only $670 million)—how could this diminutive enemy have turned back the world's military giant?

When adherents to the "win" point of view spoke of victory they usually meant one of two results: first, keeping a non-Communist government in South Vietnam; or, second, uniting under United States guidance the two Vietnams that had been divided in 1954 by the Geneva Accords. Let us consider the difficulties of achieving either goal; let us subject the "win" perspective to troubling questions and inescapable facts to demonstrate its fundamental weaknesses, as well as its inappropriateness as a justification for policy toward Central America.

To many analysts, victory would have come only through an American invasion of North Vietnam. As Secretary McNamara came to realize, the massive bombings of the Ho Chi Minh Trail, Laos, Kampuchea, and North Vietnamese cities had not forced the adversary to capitulate and had not halted the flow of arms and men to the South. The air terror hurt, of course, but it did not break the enemy, which absorbed tremendous losses. American planes dropped 6.7 billion tons of bombs on Indochina in 1965–73, or three times the total tonnage dumped on all enemy nations in the Second World War. Strategic bombing, we are told, seldom works against a non-industrialized nation. When Richard Nixon came into office in 1969, he surmised that there were only two ways he could have delivered a "knockout blow."[13] One was to bomb the North's irrigation dikes, probably killing 100,000 civilians. The other was to use tactical nuclear weapons. But he undertook neither, because he recoiled from the domestic and international uproar that surely would have met such steps, and because either action would have wrecked the détente he hoped to cultivate with the People's Republic of China.

But, even if the United States had destroyed cities, dikes, factories, and farms in the North, then what? An invasion and occupation of the North would still have been necessary. American troops would have had to slash their way across the demilitarized zone into the Democratic Republic of Vietnam, where a reserve army of nearly half a million soldiers waited. And, as a one-time war architect pointed out, "the whole population is dug in, with individual foxholes and an efficient civil defense."[14] Casualties for the United States would have been staggering. Assuming the improbable—that the invasion succeeded—the North would then have had to be occupied while Ho Chi Minh's remaining legions conducted guerrilla warfare. America's unwelcomed stay would have been bloody and prolonged, and although it might have subdued the North Vietnamese, it probably would not have controlled them. The American people would have had to be willing to fund an expensive occupation of indeterminate duration and cost against a hostile people defending their homeland against foreigners. Given all of these problems, it seems unlikely that an invasion would have worked to bring victory.

But the difficulty of invading and controlling the North did not by itself deter American leaders from pursuing such a drastic course. Their fears of Soviet or Chinese intervention probably counted as much or more. If United States forces crossed the 17th Parallel, China's leaders had stated publicly, the People's Republic of China would come to North Vietnam's defense. China had inserted troops in North Vietnam, with whom it had signed a defense treaty. Soviet Russia, North Vietnam's primary supplier, might have been dissuaded from saving its ally, but no President could gamble that both Moscow and Beijing would stand aside. American fears were anything but casual, based as they were on experience (the Korean War), intelligence data, and strategic calculations. To have invaded the North, then, a necessary action for victory, most likely would have triggered the regional war into a great power conflagration. It is difficult, of course, to see how elevating Vietnam into a world war would have led to an American victory.

Advocates of the "win" thesis also need to contend with the fact that the United States would have suffered even greater losses of international friends and prestige had it prosecuted the war more vigorously. America's European allies already thought Washington had lost its senses by investing resources in a region peripheral to the West's vital interests. NATO nations became resentful and restless, further weakening the alliance. Only four of the United States' forty worldwide allies sent troops to help in Vietnam (South Korea, Thailand, Australia, and New Zealand). And the Third World that the United States was trying so hard to woo to its Cold War positions soured even further toward Washington—measured, for example, in votes against the United States in the United Nations, restrictions on American corporations, and terrorist acts against American persons and property. When in November 1979, radical Islamic students stormed the American Embassy in Teheran, Iran, and seized Americans as hostages, one of the attackers snarled to a blindfolded captive: "We're paying you back for Vietnam."[15]

To have won the war, moreover, the United States would have had significantly to reform the Saigon government in the South—the American ally that claimed no popular following and existed at Washington's sufference. The clique of Southern leaders routinely jailed critics, fixed elections, practiced blatant corruption, played favorites with friends, refused land reform, and, as nationalists, ignored or rejected American advice. A series of conspiracies, coups, and attempted coups, and American covert actions, destabilized Vietnamese politics. America's client leader Ngo Dinh Diem polarized politics by smashing the Buddhists, leaving extreme choices: the National Liberation Front or the Saigon regime. One American-assisted plot to improve the chances for reform—the 1963 removal of Diem—only produced more instability. By 1965, George Ball concluded, South Vietnam really had an army, not a government. "I don't think we ought to take this government seriously," Ambassador Henry Cabot Lodge told a White House meeting that same year. "There is simply no one who can do anything."[16] South Vietnamese leaders came to resent American criticism; they bristled against their dependency upon American aid. The United States struggled with a dilemma: If it pressed Saigon to reform and manipulated the regime to enhance its political standing, it risked making Southern officials appear to be puppets. In so doing, the United States thus undermined its purpose of building a strong, independent government that could command popular support. In the end, American officials abandoned reform, and the weak South Vietnamese government continued to hinder victory.

Many Vietnamese refused to fight for this American-backed government. They knew about the corrupt, self-serving leaders and chose not to die for them, or for the Americans, who seemed increasingly willing to take on the responsibilities of fighting and bleeding. Marine officer Robert Muller remembered: "I served as an adviser to three separate ARVN [Army of the Republic of Vietnam] battalions, every one of which every time we were in combat, split. Not most of the time, every time!" He knew that "the writing was on the wall."[17] One-third of the men in combat units deserted each year, and this high rate, not warfare or disease, accounted for the biggest loss of manpower for America's ally. ARVN suffered the same problems that afflicted their government: poor morale, corruption, self-serving politics, and nepotism.

In short, a necessary ingredient for victory—a stable political foundation—was lacking in South Vietnam. The United States had no reliable, internal instrument for the implementation of containment. And a strong, popular, political system could not be constructed in the midst of a major war. As Ball reflected, "we failed not from military ineptitude but because there was no adequate indigenous political base on which our power could be emplaced."[18]

This question of the instability and ineffectiveness of the Southern government has led some students of the war to doubt the postwar, prescriptive analysis of Colonel Harry G. Summers (in his book *On Strategy*, 1982) and others that a defensive deployment of American forces just below the demilitarized zone along routes into the South would have stopped the movement of soldiers from the North and cut lines of supply. In military terms, defense of a 900-mile frontier against a determined and versatile enemy demonstrably adept at infiltration would have been extremely difficult at best, requiring great numbers of troops. Even if such a blocking operation had worked, the fact of tortured politics and popular discontent with the Saigon coterie would remain to insure disarray and insurgency in the South and to handicap the war effort. A different *military* strategy would not have altered *political* realities.

Another problem dogs the "win" school of thought. From the works of Colonel Summers, General Bruce Palmer (*The 25-Year War*, 1984), and Major Andrew F. Krepinevich, Jr. (*The Army and*

*Vietnam*, 1986), it seems evident that victory could never have come without a reformation of the American military establishment. The generals ordered a kill-and-destroy strategy and war of attrition that failed. Their "can-do" attitude prevented them from warning the President of the difficulty of winning. They never articulated an alternative strategy for the President, confused as they themselves were about war objectives. They kept calling for more men and bombs. The kill-and-destroy strategy alienated the Vietnamese people. The enemy was never beaten down to the point of surrender; hurt to be sure, but not conquered. As Summers has put it, the United States enjoyed tactical victory but suffered strategic defeat. A long war of attrition presumed incredible patience on the part of the American people, but they came to realize what postwar analysts have noted—the strategy was not working, and more of the same would not have insured success.

The military brass also sent soldiers into a war of insurgency after having trained them for combat in Europe or for a Korean War setting. The United States largely fought a conventional war, emphasizing massive fire-power and technology rather than light infantry. When the enemy tangled with Americans in a conventional battle, it lost; but it seldom fought that way, preferring quick attacks and skirmishes in the jungles and mountains. Although the United States trained and used some special units, they remained a secondary force, subordinate to regular combat troops. The rotation system for officers—one year in Vietnam to "punch your ticket"—broke continuity and squandered the advantages of first-hand experience.

Military leaders, like their civilian chiefs, proved woefully ignorant about Indochina—its culture, its people, its history, its record of resisting foreigners. They failed to study the long French experience or heed cautionary French advice, belittling the French as the quitters of World War II, the brutal and failed colonialists, the technologically deficient militarists (Americans had helicopters, they did not). The military also engaged in notorious waste and corruption. Reports on the massacre of Vietnamese women, children, and elderly men at My Lai in March 1968 were suppressed for twenty months. Military men played games with data. In jest, but illustrating the point, one former policymaker created a fictional setting in which a military officer informed a 1967 White House meeting on the effectiveness of the air war: American planes "knocked out 78 percent of North Vietnam's petroleum reserve; since we had knocked out 86 percent three days ago and 92 percent last week, we were doing exceptionally well."[19] Body counts were faked. Major William Lowry recalled that the "duplicity became so automatic that lower headquarters began to believe the things they were forwarding to higher headquarters. It was on paper; therefore, no matter what might have actually occurred, the paper graphs and charts became the ultimate reality."[20] Colonel John B. Keeley, who commanded an infantry battalion in the Mekong Delta in 1967–68, provided an example: "One day my battalion spent the whole day beating the bush and flushed and killed four VC [Vietcong]. Another battalion was doing the same thing and killed two VC. We sent the number four and the number two to brigade for its body count report. There, the numbers were put side by side to make 42, not the six we actually killed."[21]

The American military was also troubled by a host of discontents not unlike those afflicting American society back home. Black-white racial tension and drugs were rife. By the spring of 1971, reported American officials, 10 percent of American GIs were taking heroin; a higher percentage was smoking marijuana. In the period 1969–72, about a thousand cases of fragging were recorded—assaults on officers with the intent to kill, harm or intimidate. In the years 1965–72, some 550,000 American soldiers deserted; 570,000 American men evaded the draft. Could this military, with all of its problems, from the high-level strategist in the Pentagon to the "grunt" in the field, have won the war? Reform of the behemoth military is always difficult in peacetime; it probably would have been impossible when it most counted—during a war that was going badly.

Finally, in our discussion of why victory eluded the United States, we must consider the American people and their doubts. How could their ardent support have been aroused and maintained over a long period of time? Summers, Palmer, and others argued that a declaration of war would have mattered. It would have focused national attention on the war, stimulated the passion of the people, permitted the placement of restraints on television journalists, insured adequate mobilization, and forced dissenters to rally around the flag. This argument is questionable on several scores. Americans, as the Korean War demonstrated, become impatient with limited war, any limited war. Congress might not have declared war in 1965, when escalation began in earnest. At that time South Vietnam had hardly entered American consciousness as vital to the national interest; the Gulf of Tonkin incident of the year before lacked the characteristic of a stunning Pearl Harbor event and thus could not generate decided pro-war emotions. A declaration of war would not have changed the dismal status of the South Vietnamese government, military strategy, the continued loss of international friends, or the obstacles to invasion and occupation of the North. And for how long would Americans have tolerated governmental controls over the flow of information? Television would probably still have broadcast into homes enough gory pictures of death and destruction to convince many American citizens that their nation's conduct of the war constituted a moral disgrace. Although the credibility gap was already opened wide in 1967–68, Americans supported their government's decisions for a long time—not until 1970 did a majority call for withdrawal. By then many felt that their leaders had deceived them. Would a declaration of war have prevented deception? Would not Americans, over time, with or without a declaration of war, grow weary of the returning coffins and the higher taxes? And for how long could decision-makers persuade Americans that they should oppose Communism in Vietnam while Washington sought détente with the two great Communist states of the Soviet Union and the People's Republic of China, the latter supposedly the real enemy in Vietnam? The American people and their elected representatives did not lose their will; on the contrary, they saw well and exercised a political will that held their leaders accountable for failed policies. And surely the press cannot be faulted for simply pointing out what was obvious: the war was not being won.

Data also reveals the pale popular endorsement of the war and suggests that a declaration of war and greater mobilization would possibly have exacerbated rather than diminished protest. Besides the illegal draft evasions, legal educational and employment deferments exempted nine million men from the draft. A prolonged, *declared* war might very well have pushed up the desertion, evasion, and avoidance statistics, created a draft crisis, and set off more street demonstrations. In short, to have won the war, Washington would have had to mobilize the American people—a task that would surely have been daunted.

Why, then, did the United States lose the Vietnam War? Why could the war not have been won? First, the environment was hostile to Americans. A jungle terrain of thick grasses and bamboo, leeches, and weather conditions that alternated between hot sun and drenching rain challenged the American military. Boots and human skin rotted, and diseases flourished. "It is as if the sun and the land itself were in league with the Vietcong," recalled Marine officer Philip Caputo, "wearing us down, driving us mad, killing us."[22] The Vietnam veteran Oliver Stone's stunning 1987 movie *Platoon* revealed the inhospitable country that negated American technological superiority. Well-hidden booby traps blasted away parts of the body; snipers made every step of a "grunt" or "boonierat" precarious. The enemy was everywhere but nowhere, often burrowed into elaborate underground bunkers or melded into the population, where every Vietnamese might be a Vietcong threat. No place in Vietnam seemed secure. Many veterans later suffered "post-traumatic stress disorder," an illness of nightmares and extreme nervousness.

The United States lost the war as well, because its conduct alienated the Vietnamese people, one-third of whom became refugees. Bomb craters scarred the land; the chemical defoliant Agent Orange denuded it. Most Vietnamese were tagged "gooks," and GIs did not always distinguish between peasant farmers and the enemy. "If he's dead and Vietnamese, he's VC," became a common view.[23] Some Americans sliced off Vietnamese ears as trophies and burned to the ground villages considered friendly to the Communists. In 1967, the six thousand residents of Ben Suc, where the Vietcong had dug an elaborate tunnel network, were rounded up and sent to a refugee camp, and enemy suspects were tortured and murdered. According to the official Army history: "As the villagers and their belongings moved out, bulldozers, tankdozers, and demolition teams moved in. . . . When the village had been flattened by the engineers, . . . a large cavity was scooped out near the center of the area, filled with ten thousand pounds of explosives . . . and then set off. . . . The village of Ben Suc no longer existed."[24] As for the village of Ben Tre, "it became necessary to destroy the town to save it," remarked an American officer.[25] Just before leaving office, Defense Secretary McNamara complained about the bombing: "It's not just that it isn't preventing the supplies from getting down the trail. It's destroying the countryside in the South. It's making lasting enemies."[26] One American official later admitted that "it was as if we were trying to build a house with a bulldozer and wrecking crane."[27] The strategic hamlet program, begun in 1962, uprooted people from ancestral lands and placed them in guarded quarters surrounded by barbed wire. The South Vietnamese—a peasant people—were issued plastic identification cards. As well, the huge influx of Americans disrupted the Vietnamese economy and helped support a debasing underworld of prostitution, drugs, and the black market. Vietnam's fragile society, already bitterly divided between Catholics and Buddhists, became further fragmented. The United States in the end was destroying the very place it was trying to defend against a perceived Communist threat. No wonder Washington could not build an important, strong, popular government in the South.

The United States lost, too, because it faced a people deeply committed to their cause, who seemed to have no "breaking point."[28] When an American veteran-turned-journalist returned to Vietnam in 1984, he interviewed soldiers about why they had fought on, enduring great sacrifices and much suffering. They invariably quoted Ho Chi Minh: "Nothing is more important than independence and freedom."[29] Although probably harboring private doubts about whether such was the case, given the ruthlessness and warrior-state mentality of their Communist leaders in the 1980s, these soldiers believed in the 1960s that they were defending their nation against outsiders—as their ancestors had done for centuries against the Chinese, French, and Japanese. As General Bruce Palmer concluded, "Their will to persist was inextinguishable."[30] Vietnamese nationalism, then, ultimately forced defeat upon the United States.

General Maxwell Taylor summarized some of the reasons why Americans could not win the Vietnam War when he said that "we didn't know our ally. Secondly, we knew even less about the enemy. And, the last, most inexcusable of our mistakes, was not knowing our own people."[31]

These "mistakes," Americans were assured in the 1980s, were being avoided in El Savador where the United States sent military advisors and economic and military aid to thwart a leftist insurgency against a conservative, American-backed regime notorious for its brutalities and corruption. Cuba and the Soviet Union, Reagan charged, spawned and fed a Communist rebellion, hoping to tip a domino and challenge the United States in its own backyard. Remember, the President said, "we are the last domino."[32] Nicaragua, Reagan also believed, had become a Soviet stooge. On the other hand, many analysts did not see El Salvador as an East-West conflict and saw Nicaragua's call for Soviet help as a desperate measure to counter United States harassment. Class oppression and poverty, not Communist plots, caused people to join the rebels. With or without the Soviets, revolution would rock El Savador, where a landed elite and the military ruled. "It is not at all a question of Communist subversion," French President François Mitterrand declared, but "the people's refusal to submit to misery and humiliation."[33] Beware of another Vietnam, warned Reagan's critics. "The White House did not appreciate how rapidly El Salvador would take off in the minds of the press as a Vietnam," remarked a presidental assistant.[34]

Did the Vietnam analogy make sense? It did and it did not; but in either case it became a reference point for American intervention in Central America. Reagan seemed to think that the victory denied in Vietnam could be achieved in El Salvador. "They thought it was like rolling a drunk," Ambassador to El Salvador Robert White noted.[35] Why would Reagan and his advisers think so? There were differences between Vietnam and El Salvador that seemed favorable to a United States venture into the latter. Unlike Vietnam, which had China, El Salvador had no large Communist state at its border to provide supplies and sanctuary. El Salvador's neighbors Guatemala and Honduras were actually staunch American allies. El Salvador, moreover, was a small country about the size of Massachusetts, with only five million people, and the United States Navy had easy access to it. Logistics were far simpler than was the case for Vietnam, some 12,000 miles away. Besides this proximity, the United States also had traditional ties with El Salvador, whose military officers had trained in American schools. The President of El Salvador, José Napoléon Duarte, attended the University of Notre Dame in South Bend, Indiana. A good number of American officers spoke Spanish, whereas few had learned Vietnamese. American strategic interests in Central America were also well-defined and long-established; it thus constituted a place where Americans would no doubt fight with a staying power that was lacking in Vietnam.

The Salvadoran rebels were hardly a replica of the disciplined, tenacious Vietminh or Vietcong who became battle-hardened from years of war against the French before Americans ever became combatants. The nationalism aroused and invigorated by anti-colonialism that drove the Vietnamese, Reagan officials could reason, was not a force the Salvadoran insurgents could draw upon, for the Central American state already existed as a nation and no immediate colonial master lurked in the recent past, even if some Salvadorans considered the United States an imperial power. The Salvadoran rebels suffered internal dissension, seemed unable to gain wide popular support, and failed to control much territory. Compared to Vietnam, then, presidential advisers could tell Reagan the chances for victory in El Salvador—without the need for the introduction of American combat troops—seemed good.

The latter point loomed large, counting for another difference that Reagan officials found restraining. Americans repeatedly told pollsters and their representatives in Congress that they opposed the sending of American soldiers to Central America. Unlike 1965, when President Lyndon Johnson made the major decision to dispatch large numbers of American troops to Vietnam wtihout spawning public debate, in the 1980s Americans watched events more closely, and Congress cautioned the

President against escalation. Indeed, critics of American intervention in Central America believed that the similarities to Vietnam were stronger than the differences.

The way American officials thought about El Salvador constituted the first similarity; that is, they invoked the containment doctrine to turn back presumed Communist expansion in Central America. They once again overstated the threat, refurbished the domino theory, and intruded into a civil war in El Salvador. As in the earlier case of Vietnam, United States leaders elevated a local struggle into an international crisis by defining it as vital to American security. The "can-do" philosophy re-emerged— the "era of self-doubt is over," declared Reagan—to blind leaders to obstacles, although it seemed that the military establishment evinced reluctance this time to become involved in a land war in Central America, especially a guerrilla war.[36] Once again American policymakers arrogantly believed they had answers for other people's problems. Once again they denounced their critics as unpatriotic apologists for Communist ruthlessness. Another similarity with Vietnam: few United States allies stood with the Reagan Administration in Central America. The major nations in the region—Mexico and Venezuela—opposed the United States' emphasis on military solutions. The Contadora group (Mexico, Venezuela, Colombia, and Panama) urged negotiations on both Salvador and Nicaragua, but Washington rebuffed its repeated overtures. The Reagan Administration also flatly rejected the revolutionary Salvadorans' calls for negotiations. If the United States had violated the Geneva Accords in Vietnam, the World Court ruled in 1986 that Washington had violated international law by aiding the *contras* and should pay reparations to Nicaragua.

Another similarity between Vietnam and Central America, said opponents of intervention, lay in the unreliability and unsavory character of America's local allies. The *contra* leadership included a large number of former Somoza henchmen and National Guardsmen—probably the most hated and feared men of Nicaragua. In their war of sabotage against Nicaragua, the *contras* committed atrocities, squabbled among themselves, practiced fraud with American money, and seemed unlikely to provide a popular alternative to the Sandinistas who governed

Nicaragua. In El Salvador, although Duarte was elected, politics remained corrupt and unpredictable, and the military and wealthy remained powerful. The government appeared unable to put a stop to the ravaging death squads that silenced critics of all kinds. Yet another similarity flowed from these ties to local leaders: a credibility gap. If Americans came to doubt that officials were telling them the truth in the Vietnam era, they also came to question Washington's exaggerated rhetoric about Central America. Reagan's depiction of the *contras* as "freedom fighters" and his assurances that "democracy" was emerging and "human rights" were improving in El Salvador seemed ludicrous. The official *White Paper* of 1981 on El Salvador was so riddled with errors and unsubstantiated generalizations about a Communist plot that the administration soon shelved it.

As of this writing, the outcome of increased United States intervention in Central America remains uncertain. Of course, the scenario will not follow each step of the Vietnam experience. Reagan has vowed never to send American troops to Central America. But if the critics are right, the United States will stumble in Central America for some of the same reasons it lost in Vietnam. The regions contrast sharply, but American assumptions have not changed and local conditions promise again to determine the outcome. Once again the exaggerated image of a relentless Communist threat seems to compel the United States to intervene in civil wars for which there are seldom outside answers. The past directs—and continues to mislead. "Here we are again," boomed Republican Senator Mark O. Hatfield of Oregon, "old men creating a monster for young men to destroy."[37]

# Notes

## Chapter 1

This chapter, under a slightly different title and with fuller documentation, first appeared in the *American Historical Review*, LXXV (April 1970), 1046–64. Les K. Adler was co-author. This essay aroused debate, as seen in spirited exchanges in the *Review*, LXXV (December 1970), 2155–64; LXXVI (April 1971), 575–80; and LXXVI (June 1971), 856–58. Although substantially the same as the published article, this shorter chapter reflects rewriting to account for criticisms and subsequent scholarship, including Daniel M. Smith, "Authoritarianism and American Policy Makers in Two World Wars," *Pacific Historical Review*, XLIII (August 1974), 303–23 and Thomas R. Maddux, "Red Fascism, Brown Bolshevism: The American Image of Totalitarianism in the 1930s," *The Historian*, XL (Nov. 1977), 85–103.

1. Herbert L. Matthews, "Fascism Is Not Dead . . .," *Nation's Business*, XXXIV (Dec. 1946), 40.
2. *Public Papers of the President of the United States, Harry S. Truman, 1947* (Washington, D.C., 1963), 238 (hereafter cited as *Public Papers*).
3. Matthews, "Fascism," 40.
4. See, for example, J. Edgar Hoover, "Red Fascism in the United States Today," *American Magazine*, CXLIII (Feb. 1947), 24; Jack B. Tenney, *Red Fascism* (Los Angeles, 1947); Howard K. Smith, *The State of Europe* (New York, 1949), 67; Norman Thomas, "Which Way

America—Fascism, Communism, Socialism, or Democracy?" *Town Meeting Bulletin*, XIII (March 16, 1948), 19–20.

5. John P. Diggins, "Flirtation with Fascism: American Pragmatic Liberals and Mussolini's Italy," *American Historical Review*, LXXI (Jan. 1966), 499.

6. Herbert J. Spiro, "Totalitarianism," in David L. Sills, ed., *International Encyclopedia of the Social Sciences* 17 vols. (New York, 1968), XVI, 112.

7. George F. Kennan, "Totalitarianism in the Modern World," in Carl J. Friedrich, ed., *Totalitarianism* (Cambridge, Mass., 1954), 19.

8. Eugene Lyons, *Assignment in Utopia* (New York, 1937), 611, 621–22.

9. Hannah Arendt, *The Origins of Totalitarianism* (Cleveland, 1958), 347.

10. *New York Times*, Sept. 18, 1939.

11. Quoted in Ronald E. Magden, "Attitudes of the American Religious Press toward Soviet Russia, 1939–1941" (Ph.D. diss., Univ. of Washington, 1964), 38.

12. Quoted in *ibid.*, 79.

13. *Wall Street Journal*, June 25, 1941.

14. *New York Times*, July 15 and June 24, 1941.

15. Quoted in Paul Willen, "Who 'Collaborated' with Russia?," *Antioch Review*, XIV (Sept. 1954), 262.

16. Walter Millis, ed., *The Forrestal Diaries* (New York, 1951), 47.

17. U.S. Department of State, *Foreign Relations of the United States: The Conference of Berlin, 1945*, I, 274. The *Foreign Relations* series is published by the Department of State in Washington, D.C. in multiple volumes for each year, although some volumes are devoted to particular conferences. Hereafter cited as *Foreign Relations*.

18. Millis, *Forrestal Diaries*, 128.

19. Memorandum by Louis E. Wyman, June, 1946, Charles Tobey Papers, Box 49, Dartmouth College Library, Hanover, N.H.

20. *Department of State Bulletin*, XII (June 17, 1945), 1097–98.

21. John R. Deane, *Strange Alliance* (New York, 1947), 4, 216, 219, 306.

22. U.S. Congress, Senate Committee on Foreign Relations, *European Recovery Program* (Hearings, Washington, D.C., 1948), 1191.

23. Address, Feb. 13, 1946, Bernard Baruch Papers, Princeton Univ. Library, Princeton, N.J.

24. *Department of State Bulletin*, XX (Feb. 27, 1949), 248.

25. Address, April 4, 1947, Arthur Bliss Lane Papers, Yale Univ. Library, New Haven, Conn.

26. *Foreign Relations, Berlin, 1945*, I, 272.

27. "Have Britain and America Any Reason to Fear Russia?" *Town Meeting Bulletin*, XI (March 21, 1946), 5.
28. Speech, "Save Human Freedom," Feb. 16, 1947, J. Howard Mc-Grath Papers, Harry S. Truman Library, Independence, Mo. (hereafter cited as HSTL).
29. George F. Kennan, "Overdue Changes in Our Foreign Policy," *Harper's*, CCXIII (Aug. 1956), 27–33.
30. George F. Kennan, *Memoirs, 1925–1950* (Boston, 1967), 403.
31. Arthur H. Vandenberg, Jr., *The Private Papers of Senator Vandenberg* (Boston, 1952), 161.
32. *Ibid.*, 342.
33. Millis, *Forrestal Diaries*, 96.
34. *Barron's*, XXVI (Sept. 23, 1946), 1.
35. Quoted in Walter LaFeber, *America, Russia, and the Cold War, 1945–1984*, 5th ed. (New York, 1985), 123.
36. *Public Papers, Truman, 1945* (Washington, D.C., 1961), 433.
37. Quoted in D. N. Pritt, *The State Department and the Cold War* (New York, 1948), 16.
38. American Opinion Report, Feb. 9, 1948, Department of State Records, National Archives, Washington, D.C. (hereafter cited as State Records).
39. *New York Times*, Jan. 22, 23, 1948.
40. Kennan, *Memoirs, 1925–1950*, 268.
41. Quoted in Peter G. Filene, ed., *American Views of Soviet Russia, 1917–1965* (Homewood, Ill., 1968), 242.
42. Wallace to Truman, July 23, 1946, Clark Clifford Papers, HSTL.
43. Kennan, "Overdue Changes," 28.
44. *Reader's Digest*, LV (Sept. 1949), 156.
45. Kennan, "Totalitarianism," 19–20.

# Chapter 2

This chapter is my revised, abbreviated version of "The Quest for Peace and Prosperity: International Trade, Communism, and the Marshall Plan," first published in Barton J. Bernstein, ed., *Politics and Policies of the Truman Administration* (Chicago, Quadrangle Books, 1970), 78–112.

1. *Foreign Relations, 1946*, VI, 708.
2. James A. Stilwell, *Department of State Bulletin*, XIV (May 19, 1946), 831.
3. U.S. Department of State, *Development of the Foreign Reconstruction*

*Policy of the United States, March–July, 1947* (Washington, D.C., 1947), 13.

4. U.S. Economic Cooperation Administration, *European Recovery Program: Western Germany, Country Study* (Washington, D.C., 1949), 1.
5. U.S. Congress, Senate Committee on Foreign Relations, *Emergency Recovery Program* (Hearings, Washington, D.C., 1948), 245.
6. "First Staff Draft of the Council of Economic Advisers Report on the Foreign Aid Program," Sept. 10, 1947, "Scope of Recommendation" folder, President's Committee on Foreign Aid Records, HSTL.
7. *Foreign Relations, 1947*, III, 230.
8. *Department of State Bulletin*, XVI (June 15, 1947), 1160.
9. *Public Papers, Truman, 1946* (Washington, D.C., 1962), 354.
10. *Ibid., 1947*, 167.
11. *Department of State Bulletin*, XII (May 27, 1945), 979.
12. Paul G. Hoffman, "The Survival of Free Enterprise," *Harvard Business Review*, XXV (Autumn 1946), 24.
13. *World Trade*, XII (July 1946), 2.
14. *Nation's Business*, XXXIII (Aug. 1945), 50.
15. *Journal of Commerce*, Feb. 24, 1947.
16. Truman in U.S. Congress, Senate Committee on Foreign Relations, *European Recovery Program: Basic Documents and Background* (Washington, D.C., 1947), 67; Krug in Senate, *European Recovery Program*, 354.
17. U.S. Department of State, *Peace, Freedom, and World Trade* (Washington, D.C., 1947), 5.
18. U.S. Department of State, *Problems of United States Foreign Economic Policy* (Washington, D.C., 1947), 3.
19. Snyder to the President, "Comments on Draft of the Economic Report of the President," n.d., OF 396, Harry S. Truman Papers, HSTL.
20. Theodore A. Sumberg, "The Government's Role in Export Trade," *Harvard Business Review*, XXIII (Winter 1945), 167.
21. Committee for Economic Development, *International Trade, Foreign Investment, and Domestic Employment Including the Bretton Works Proposal* (New York, 1945), 10.
22. B. C. Forbes, ed., *America's Fifty Foremost Business Leaders* (New York, 1948), 226.
23. *Department of State Bulletin*, XXI (Sept. 12, 1949), 401.
24. James Forrestal to Paul C. Smith, March 19, 1947, Box 91, James Forrestal Papers, Princeton Univ. Library.
25. Telephone Conversation with Paul Shields, March 20, 1947, Forrestal Papers.

26. "Draft of March 10, 1947," Box 17, George Elsey Papers, HSTL.
27. *Journal of Commerce*, Oct. 31, 1946.
28. *Ibid.*, June 3, 1947.
29. *Department of State Bulletin*, XVII (Nov. 16, 1947), 933.
30. Memorandum, Turner to Steelman, March 5, 1948, OF 396, Truman Papers.
31. Senate, *European Recovery Program*, 58.
32. U.S. National Advisory Council on International Monetary and Financial Problems, *Statement of the Foreign Loan Policy of the United States Government* (Washington, D.C., 1946), 5.
33. Quoted in Thomas G. Paterson, *Soviet-American Confrontation: Postwar Reconstruction and the Origins of the Cold War* (Baltimore, Md., 1973), 215.
34. Vera Micheles Dean, "Economic Trends in Eastern Europe—II," *Foreign Policy Reports*, XXIV (April 15, 1948), 38.
35. Quoted in James Riddleberger (State Department) to Ambassador Laurence Steinhardt, Oct. 3, 1946, Box 51, Laurence Steinhardt Papers, Library of Congress, Washington, D.C.
36. Kennan, *Memoirs, 1925–1950*, 342; Clayton in *Foreign Relations, 1947*, III, 232 (italics in original).
37. *Journal of Commerce*, Jan. 30, 1946.
38. Quoted in Henry W. Berger, "A Conservative Critique of Containment: Senator Taft on the Early Cold War Program," in David Horowitz, ed., *Containment and Revolution* (Boston, 1967), 131–32.
39. Arthur Besse in *Journal of Commerce*, March 12, 1947.

## Chapter 3

This chapter was first composed as a lecture and delivered in 1984 to the Truman Centennial Celebration, University of Missouri at Kansas City. For their discussions of exaggerations of the Soviet threat, this chapter is indebted to Melvyn P. Leffler, "The American Conception of National Security and the Beginnings of the Cold War, 1945–1948," *American Historical Review*, LXXXIX (April 1984), 346–81; Matthew A. Evangelista, "Stalin's Postwar Army Reappraised," *International Security*, VII (Winter 1982–83), 110–38; Samuel F. Wells, Jr., "Sounding the Tocsin: NSC-68 and the Soviet Threat," *International Security*, IV (Fall 1979), 116–58; and David A. Rosenberg, "American Atomic Strategy and the Hydrogen Bomb Decision," *Journal of American History*, LXVI (June 1979), 62–87.

1. Quoted in *New York Times*, Oct. 27, 1976.
2. Jimmy Carter, *Keeping Faith: Memoirs of a President* (New York, 1982), 65–66.
3. Robert K. Murray and Tim H. Blessing, "The Presidential Performance Study: A Progress Report," *Journal of American History*, LXX (Dec. 1983), 535–55.
4. Quoted in Margaret Truman, *Harry S. Truman* (New York, 1973), 207–12.
5. Robert H. Ferrell, ed., *Off the Record: The Private Papers of Harry S. Truman* (New York, 1980), 16.
6. Thomas M. Campbell and George C. Herring, eds., *The Diaries of Edward R. Stettinius, Jr., 1943–1946* (New York, 1975), 325.
7. Quoted in Truman, *Truman*, 358.
8. Quoted in Terry H. Anderson, *The United States, Great Britain, and the Cold War, 1944–1947* (Columbia, Mo., 1981), 54.
9. Quoted in Truman, *Truman*, 213.
10. Robert H. Ferrell, ed., *Dear Bess: The Letters from Harry to Bess Truman, 1910–1959* (New York, 1983), 518.
11. Harry S. Truman, *Memoirs*, 2 vols. (Garden City, N.Y., 1955–56), I, 341.
12. Ferrell, *Dear Bess*, 519–20.
13. Quoted in Barton J. Bernstein, "American Foreign Policy and the Origins of the Cold War," in Barton J. Bernstein, ed., *Politics and Policies of the Truman Administration* (Chicago, 1970), 32.
14. Quoted in John L. Gaddis, *The United States and the Origins of the Cold War, 1941–1947* (New York, 1972), 205.
15. W. Averell Harriman and Elie Abel, *Special Envoy to Churchill and Stalin, 1941–1946* (New York, 1975), 452.
16. Quoted in Truman, *Truman*, 269.
17. Diary, Oct. 17, 1947, Box 2, Felix Frankfurter Papers, Library of Congress.
18. Quoted in Thomas G. Paterson, *On Every Front: The Making of the Cold War* (New York, 1979), 69.
19. Irving Brant quoted in Richard C. Lukacs, *Bitter Legacy: Polish-American Relations in the Wake of World War II* (Lexington, Ky., 1982), 33.
20. Lucius D. Clay, *Decision in Germany* (Garden City, N.Y., 1950), 21.
21. Trygve Lie, *In the Cause of Peace: Seven Years with the United Nations* (New York, 1954), 223.
22. Dean Acheson, transcript of proceedings, "American Foreign Policy," June 4, 1947, Box 93, United States Mission to the United Nations Records, National Archives.

23. "Remarks by the President to the Group Meeting at the White House," draft, Oct. 27, 1947, OF 426, Truman Papers.
24. Quoted in Robert L. Messer, *The End of an Alliance: James F. Byrnes, Roosevelt, Truman, and the Origins of the Cold War* (Chapel Hill, N.C., 1982), 128.
25. Henry L. Stimson and McGeorge Bundy, *On Active Service in Peace and War* (New York, 1948), 644.
26. Quoted in David MacIsaac, "The Air Force and Strategic Thought, 1945–1951," *Working Paper No. 8* (Washington, D.C., International Security Studies Program, Woodrow Wilson Center, 1979), 6, 10.
27. Telephone conversation, Senator Claude Pepper and James Forrestal, March 24, 1947, Box 91, Forrestal Papers.
28. *Foreign Relations, 1950,* I, 252.
29. Truman quoted in Alfred Steinberg, *The Man from Missouri: The Life and Times of Harry S. Truman* (New York, 1962), 259; Acheson, *New York Times,* Oct. 31, 1971.
30. Mr. "X" [George F. Kennan], "The Sources of Soviet Conduct," *Foreign Affairs,* XXV (July 1947), 581.
31. Melvyn P. Leffler, "Adherence to Agreements: Yalta and the Experiences of the Early Cold War," *International Security,* XI (Summer 1986), 94.
32. Quoted in Barton Gellman, *Contending with Kennan: Toward a Philosophy of American Power* (New York, 1984), 97.
33. U.S. Congress, Senate Committee of Foreign Relations, *North Atlantic Treaty* (Hearings, Washington D.C., 1949), 213.
34. Dean Acheson in Princeton Seminar transcript, Oct. 10–11, 1953, Box 65, Dean Acheson Papers, HSTL.
35. Minutes, 35th Annual Meeting, United States Chamber of Commerce, April 28, 1947, Chamber of Commerce Library, Washington, D.C.
36. Quoted in Robert L. Messer, "Paths Not Taken: The United States Department of State and Alternatives to Containment, 1945–1946," *Diplomatic History,* I (Fall 1977), 302.
37. *Public Papers, Truman, 1947,* 178.
38. Minutes of Subcommittee on Foreign Policy, Information Meeting of March 6, 1947, Box 88, State-War-Navy Coordinating Committee (SWNCC) Records, National Archives.
39. Walter Lippmann, *The Cold War: A Study in U.S. Foreign Policy* (New York, 1947), 11–12, 14, 38, 50–52.
40. Quoted in Wells, "Sounding the Tocsin," 128.
41. *Foreign Relations, 1950,* I, 237, 244, 263, 264.
42. *Ibid.,* 222.

## Chapter 4

This chapter is a revised version of "If Europe, Why Not China?" The Containment Doctrine, 1947–1949," *Prologue*, XIII (Spring 1981), 18–38.

1. Interview with Dean Acheson, Feb. 17, 1955, Box 1, Memoirs File, Post-Presidential Files, Truman Papers.
2. *Press Conferences of the Secretary of State*, March 18, 1947 (microfilm, Wilmington, Del: Scholarly Resources, Inc.)
3. U.S. Department of State, *United States Relations with China* (Washington, D.C., 1949), 582.
4. Quoted in Ross Y. Koen, *The China Lobby in American Politics* (New York, 1974), 58.
5. Vandenberg, *Private Papers*, 523.
6. *Ibid.*
7. Quoted in James Fetzer, "Senator Vandenberg and the American Commitment to China, 1945–1950," *The Historian*, XXXVI (Feb. 1974), 289.
8. U.S. Congress, Senate Committee on Appropriations, *Third Supplemental Appropriation Bill* (Hearings, Washington, D.C., 1947), 133.
9. *Congressional Record*, XCV (Sept. 9, 1949), 12755.
10. *Ibid.*, 12758.
11. Cabinet Meeting of Aug. 2, 1946, Notes on Cabinet Meetings, Matthew J. Connelly Papers, HSTL.
12. SWNCC Minutes, June 26, 1947, Box 19, Interdepartmental and Intradepartmental Committee Records, State Records, National Archives.
13. *Selected Works of Mao Tse-tung*, 5 vols. (Peking, 1961–77), IV, 433.
14. O. Edmund Clubb, *The Witness and I* (New York, 1974), 76.
15. Herbert Feis, *The China Tangle* (New York, 1965), 272.
16. Truman, *Memoirs*, II, 76.
17. *Foreign Relations, 1946*, VI, 704.
18. Telegram, Acheson to Certain American Missions, June 19, 1946, Box 2481, Athens Post Files, Foreign Service Post Records, State Records, National Archives; Cabinet Meeting of Aug. 2, 1946, Notes, Connelly Papers.
19. *Foreign Relations, 1947*, VII, 7.
20. *Ibid., 1946*, X, 24.
21. *Ibid., 1947*, VII, 838–39.
22. *Ibid.*, 849.
23. Minutes of Committee of Three, Nov. 3, 1947, State Records, National Archives.
24. *Foreign Relations, 1949*, VIII, 368.

25. *Ibid.*, *1948*, VIII, 147.
26. U.S. Department of State, *U.S. Relations with China*, xvi–xvii.
27. "Possible Developments in China: Summary," CIA-ORE 27–48, Nov. 3, 1948, Records of the Army Staff, National Archives.
28. U.S. Department of State, *U.S. Relations with China*, xvi.
29. *Foreign Relations, 1949*, IX, 466.
30. Quoted in David McLean, "American Nationalism, the China Myth, and the Truman Doctrine: The Question of Accommodation with Peking, 1949–1950," *Diplomatic History*, X (Winter 1986), 32.
31. Secretary's Weekly Summary, June 23, 1947, State Records, National Archives.
32. *Foreign Relations, 1947*, I, 745.
33. Quoted in Ernest R. May, *The Truman Administration and China, 1945–1949* (Philadelphia, 1975), 83.
34. *Foreign Relations, 1948*, VIII, 78.
35. U.S. Department of State, *U.S. Relations with China*, 281.
36. U.S. Congress, House Committee on International Relations, *United States Policy in the Far East: Executive Sessions, Historical Series* (Washington D.C., 1976), VII, Part 1, 182.
37. *Ibid.*, 179.
38. Cabinet Meeting of March 7, 1947, Notes, Connelly Papers.
39. U.S. Congress, Senate Committee on Foreign Relations, "State Department Nominations," June 21, 1949, Senate Records, National Archives.
40. Transcript, "A Conversation with Dean Acheson," by Eric Sevareid, Sept. 28, 1969, CBS Television, 2.
41. *Foreign Relations, 1948*, V, Part 1, 185–86.
42. U.S. Congress, House Committee on International Relations, *Military Assistance Programs: Executive Sessions, Historical Series* (Washington D.C., 1976), V, Part 1, 228.
43. U.S. Department of State, "Round Table Discussion on American Policy Toward China," Oct. 6–7, Box 174, President's Secretary's File, Truman Papers.
44. *Department of State Bulletin*, XXII (Jan. 23, 1950), 116–17.
45. U.S. Congress, House Committee on Foreign Affairs, *Assistance to Greece and Turkey* (Hearings, Washington, D.C., 1947), 18, 49.
46. *New York Times*, Jan. 8, 1950.
47. Walter Lippmann to Alan G. Kirk, April 26, 1950, Box 82, Walter Lippmann Papers, Yale Univ. Library.
48. Cabinet Meeting of Nov. 26, 1948, Notes, Connelly Papers.
49. U.S. Congress, House Committee on International Relations, *For-*

*eign Economic Assistance Programs: Executive Sessions, Historical Series* (Washington, D.C., 1976), IV, Part 2, 66.

50. *Public Papers, Truman, 1949* (Washington, D.C., 1964), 408–9.
51. George F. Kennan to Secretary George C. Marshall, June 18, 1949, Box 13, Country and Area Files, Policy Planning Staff Records, State Records, National Archives.
52. *Public Papers, Truman, 1948* (Washington, D.C., 1964), 408.
53. Quoted in Daniel Yergin, *Shattered Peace* (Boston, 1977), 441.
54. U.S. Department of State, *U.S. Relations with China*, 606.
55. Policy Planning Staff, "Résumé of World Situation," 1947, PPS/13, Box 3, Policy Planning Staff Records.
56. Quoted in Truman, *Truman*, 412.
57. U.S. Congress, Senate Committee on Foreign Relations, *The United States and Communist China in 1949 and 1950: The Question of Reapprochement and Recognition*, by Robert M. Blum (Washington, D.C., 1973), 4.
58. *Foreign Relations, 1949*, VIII, 388.
59. *Ibid., 1949*, IX, 157.
60. *Ibid., 1950*, VII, 1368.

# Chapter 5

This chapter originally appeared with full documentation in *Diplomatic History*, III (Winter 1979), 1–18, under the title "Presidential Foreign Policy, Public Opinion, and Congress: The Truman Years." Copyright © by Scholarly Resources, Inc. Used by permission.

1. George Elsey Oral History, HSTL.
2. Clinton Rossiter, *The American Presidency* (New York, 1956), 52.
3. Bernard C. Cohen, "The Relationship Between Public Opinion and Foreign Policy Makers," in Melvin Small, ed., *Public Opinion and Historians* (Detroit, 1970), 79.
4. Rossiter, *American Presidency*, 133.
5. Quoted in Truman, *Truman*, 356.
6. Quoted in *ibid.*, 353.
7. James Reston, "The Number One Voice," in Lester Markel, et al. *Public Opinion and Foreign Policy* (New York, 1949), 66.
8. Quoted in Richard E. Neustadt, *Presidential Power* (New York, 1960), 50.
9. Quoted in Carl L. Becker, *How New Will the Better World Be?* (New York, 1944), 25.
10. Quoted in Geoffrey Perrett, *Days of Sadness, Years of Triumph* (Baltimore, 1974), 418.

11. Harriman and Abel, *Special Envoy*, 531; Acheson, *Department of State Bulletin*, XIV (June 16, 1946), 1045.
12. Wayne Morse to Andrew Comrie, Oct. 22, 1945, Box S-4, Wayne Morse Papers, Univ. of Oregon Library, Eugene, Ore.
13. Quoted in Robert A. Divine, *Second Chance* (New York, 1967), 280.
14. Vandenberg, *Private Papers*, 1.
15. *Congressional Record*, XCI (Jan. 10, 1945), 166.
16. Becker, *How New*, 43.
17. Quoted in Justus D. Doenecke, "The Strange Career of American Isolationism, 1944–1954," *Peace and Change*, III (Summer-Fall 1975), 80.
18. Quoted in Martin Kriesberg, "Dark Areas of Ignorance," in Markel, *Public Opinion*, 54.
19. James N. Rosenau, *Public Opinion and Foreign Policy* (New York, 1961), 35.
20. Quoted in Kensuke Yanagiya, "The Renewal of ERP: A Case Study" (unpublished paper, Public Affairs 520-D, Woodrow Wilson School of Public and International Affairs, Princeton Univ. 1952), 20.
21. Quoted in Lester Markel, "Opinion—A Neglected Instrument," in Markel, *Public Opinion*, 31.
22. Gabriel A. Almond, *The American People and Foreign Policy* (New York, 1950), 138.
23. Chadwick F. Alger, "The External Bureaucracy in United States Foreign Affairs," *Administrative Science Quarterly*, VII (June 1962), 50–78.
24. Neustadt, *Presidential Power*, 49.
25. Truman to John H. Folger, April 19, 1947, Box 141, President's Secretary's File, Truman Papers.
26. Clark Clifford, "Memorandum for the President," Nov. 19, 1947, Box 21, Clifford Papers.
27. Thomas A. Bailey, *Man in the Street* (New York, 1948), 13.
28. Quoted in Yergin, *Shattered Peace*, 6.
29. Charles Frankel, quoted in Bernard C. Cohen, *The Public's Impact on Foreign Policy* (Boston, 1973), 178–79.
30. *Foreign Relations, 1945*, V, 257.
31. Memorandum of Conversation, May 30, 1945, Vol. 7, Conversations, Joseph Grew Papers, Harvard Univ. Library.
32. Truman, *Memoirs*, II, 177.
33. *Foreign Relations: Conferences at Cairo and Teheran*, 594.
34. Quoted in Floyd M. Riddick, "The First Session of the Eightieth Congress," *American Political Science Review*, XLII (Aug. 1948), 683.

35. Cabinet Meeting of June 9, 1949, Notes, Connelly Papers.
36. Truman, *Memoirs*, II, 454.
37. Dean Acheson, *A Citizen Looks at Congress* (New York, 1957), 53.
38. "Should Truman's Greek and Turkish Policy Be Adopted: A Radio Discussion," *University of Chicago Roundtable*, April 20, 1947, 7, 2.
39. U.S. Congress, Senate Committee on Foreign Relations, *Legislative Origins of the Truman Doctrine: Executive Sessions, Historical Series* (Washington, D.C., 1973), 142.
40. Arthur Vandenberg to Bruce Barton, March 24, 1947, Box 2, Arthur Vandenberg Papers, Univ. of Michigan Library, Ann Arbor, Mich.
41. Quoted in Yergin, *Shattered Peace*, 172.
42. Vandenberg, *Private Papers*, 550–51 (emphasis added).
43. Gaddis Smith, *Dean Acheson*, (New York, 1972), 407.
44. James A. Robinson, *Congress and Foreign Policy-Making*, rev. ed. (Homewood, Ill., 1967), 46.
45. Truman, *Memoirs*, II, 211.
46. Quoted in Theodore A. Wilson and Richard D. McKinzie, "White House Versus Congress: Conflict or Collusion? The Marshall Plan as a Case Study" (unpublished paper, 1973), 2.
47. Acheson, *A Citizen Looks*, 75.
48. Princeton Seminar transcript, Feb. 13–14, 1954, Box 66, Acheson Papers, HSTL.
49. Quoted in Doris A. Graber, *Public Opinion, the President, and Foreign Policy* (New York, 1968), 340.

# Chapter 6

This chapter constitutes abridged and rewritten portions of my contributions to my edited volume, *Cold War Critics: Alternatives to American Foreign Policy in the Truman Years*, published in 1971 by Quadrangle Books of Chicago.

1. Quoted in H. C. Peterson and Gilbert Fite, *Opponents of War, 1917–1918* (Seattle, 1968; c. 1957), 14.
2. Vera M. Dean, "Is Russia Alone to Blame?" *Foreign Policy Reports*, XXV (March 8, 1946).
3. Quoted in Petition, Nov. 13, 1946, UNRRA Files, Herbert Lehman Papers, Columbia Univ. Library, New York.
4. Elliott Roosevelt, *As He Saw It* (New York, 1946), xviii.
5. Clark Clifford, "Memorandum for the President," Nov. 19, 1947, Box 21, Clifford Papers.
6. Quoted in Athan Theoharis, "The Rhetoric of Politics: Foreign

Policy, International Security, and Domestic Politics in the Truman Era, 1945–1950," in Bernstein, *Politics and Policies of the Truman Administration*, 215.

7. Richard Scandrett to Phyllis Auty, Oct. 25, 1948, Vol. 20-A, Richard Scandrett Papers, Cornell Univ. Library, Ithaca, N.Y.
8. Vandenberg, *Private Papers*, 498.
9. Clark Clifford, "Memorandum for the President," Nov. 19, 1947, Box 21, Clifford Papers.
10. Quoted in Robert Sherrill, *Gothic Politics in the Deep South: Stars of the New Confederacy* (New York, 1968), 148–49.
11. Truman, *Memoirs*, II, 185.
12. Quoted in Alexander R. Stoesen, "The Senatorial Career of Claude D. Pepper" (unpublished Ph.D. diss., Univ. of North Carolina, 1965), 215–16.
13. *Ibid.*, 241.
14. Speech to Executive Club of Chicago, Feb. 20, 1948, Box 122, Claude Pepper Papers, Federal Records Center, Suitland, Md.
15. "Conversation Between Stalin and Members of the House Special Committee on Post-War Economic Policy and Planning," Sept. 14, 1945, Box 47; "Conversation with Stalin," Box 46, Pepper Papers.
16. Edward R. Stettinius, Jr., *Roosevelt and the Russians: The Yalta Conference* (Garden City, N.Y., 1949), 121.
17. *Foreign Relations, Conference at Malta and Yalta, 1945*, 313.
18. Quoted in Thomas G. Paterson, *Soviet-American Confrontation: Postwar Reconstruction and the Origins of the Cold War* (Baltimore, Md., 1973), 33.
19. Albert Z. Carr, *Truman, Stalin, and Peace* (Garden City, N.Y., 1950), 41.
20. "Paper on German Reparations, Revised, November 17, 1947," Edwin Pauley Papers (in his possession).
21. Speech to Democratic Club of Baltimore, Aug. 21, 1946, Box 44, Pepper Papers.
22. Claude Pepper, "A Program for Peace," *The New Republic*, CXLIV (April 8, 1946), 470.
23. Claude Pepper, "Things We Forget About Russia," *The Churchmen*, June 1, 1946, 14.
24. *Congressional Record*, XCII (April 4, 1946), 3087.
25. "Soviet Agriculture," n.d., Box 44, Pepper Papers.
26. *Congressional Record*, XCII (March 20, 1946), 2466.
27. Henry A. Wallace to the President, July 23, 1946, in *The New Republic*, CXV (Sept. 30, 1946), 402.
28. *Foreign Relations, Yalta*, 903.

29. *Congressional Record,* XCIII (April 17, 1947), 3603.
30. *Ibid.,* April 10, 1947, 3283, 3287; March 25, 1947, 2527.
31. "Labor Day Speech," Sept. 4, 1950, Box 124, Pepper Papers.
32. Quoted in Stoesen, "Senatorial Career," 230.
33. In 1962, Claude Pepper was elected congressman from Florida, and, as of this writing, he still sits there as a vocal and respected advocate of federal programs for elderly citizens.

## Chapter 7

This chapter is a revision based upon two previously published essays: "The Search for Meaning: George F. Kennan and American Foreign Policy," in Frank Merli and Theodore Wilson, eds., *Makers of American Diplomacy* (New York, Scribner's, 1974), 553–58, and "George F. Kennan," in the *International Encyclopedia of the Social Sciences—Biographical Supplement,* Vol. XIX (New York, The Free Press, 1979), 375–81.

1. Kennan, *Memoirs, 1925–1950,* 293–94.
2. *Foreign Relations, 1946,* VI, 696–709.
3. Kennan, *Memoirs, 1925–1950,* 294.
4. George F. Kennan, "History and Diplomacy as Viewed by a Diplomatist," *Review of Politics,* XVIII (April 1956), 176.
5. Kennan, *Memoirs, 1925–1950,* 9.
6. *Ibid.,* 264.
7. Quoted from the George F. Kennan Papers, Princeton Univ. Library, in Michael Rogan, "George F. Kennan and Communism: The Functioning of an Image" (unpublished seminar paper, Univ. of Connecticut, 1971).
8. George F. Kennan, "Lectures on Foreign Policy," *Illinois Law Review,* XIV (Jan.–Feb. 1951), 738.
9. Kennan, *Memoirs, 1925–1950,* 7.
10. *Ibid.,* 11, 15.
11. Quoted in C. Ben Wright, "George F. Kennan, Scholar and Diplomat: 1926–1946" (unpublished Ph.D. diss., Univ. of Wisconsin, 1972), 28.
12. Kennan, *Memoirs, 1925–1950,* 57.
13. *Ibid.,* 67; George F. Kennan, "Germany-Soviet Union." *Foreign Service Journal,* XXVII (Feb. 1950), 25; Wright, "Kennan," 96.
14. Quoted in Wright, "Kennan," 84–85.
15. Kennan, *Memoirs, 1925–1950,* 68, 69.
16. "Russia," lecture at Foreign Service School, May 20, 1938, Kennan Papers.

17. Charles W. Thayer, *Diplomat* (New York, 1959), 179.
18. Kennan, *Memoirs, 1925–1950,* 232, 253.
19. *Ibid.,* 238.
20. *Foreign Relations, 1945,* V, 901–8.
21. Kennan, *Memoirs, 1925–1950,* 545.
22. Quoted in Wright, "Kennan," 228, 336; Kennan, *Memoirs, 1925–1950,* 178, 258.
23. Kennan, *Memoirs, 1925–1950,* 295.
24. *Department of State Bulletin,* XVI (May 11, 1947), 924.
25. *Foreign Relations, 1947,* III, 223–30.
26. Mr. "X," "The Sources," 575, 576, 581, 582.
27. Lippmann, *The Cold War,* 18.
28. Mr. "X," "Sources," 574.
29. Quoted in John Lewis Gaddis, *Strategies of Containment* (New York, 1982), 40.
30. Speech, "Requirements of National Security," Jan. 23, 1947, Kennan Papers; *New York Times,* Aug. 23, 1950.
31. Quoted in Gellman, *Contending with Kennan,* 127.
32. Kennan, *Memoirs, 1925–1950,* 356.
33. Quoted in Wells, "Sounding," 122.
34. Noble Frankland and Royal Institute of International Affairs, eds., *Documents on International Affairs, 1957* (London, 1960), 157.
35. George F. Kennan, *Russia, the Atom, and the West* (New York, 1958), 14, 38, 62, 66, 92.
36. *Congressional Record,* CIV (Jan. 16, 1958), 553.
37. Dean Acheson, "The Illusion of Disengagement," *Foreign Affairs,* XXXVI (April 1958), 371–82.
38. "Transcript of a Round Table Discussion on the Theme of *Russia, the Atom, and the West,*" Paris, Jan. 18, 1958, Kennan Papers.
39. George F. Kennan Oral History, John F. Kennedy Library, Boston, Mass. (hereafter cited as JFKL).
40. George F. Kennan, *Memoirs, 1950–1963* (Boston, 1972), 306.
41. *Ibid.,* 323.
42. U.S. Congress, Senate Committee on Foreign Relations, *Supplemental Foreign Assistance, Fiscal Year 1966—Vietnam* (Hearings, Washington, D.C., 1966), 386.
43. Transcript, "Moyers Journal: Conversation with George Kennan," Jan. 30, 1973, WGBH-TV, Boston, Mass.
44. George F. Kennan, "Fresh Look at our China Policy," *New York Times Magazine,* Nov. 22, 1964, 21.
45. George F. Kennan, *The Cloud of Danger* (Boston, 1977), 26.
46. *Ibid.,* 70.

47. *New York Times*, Feb. 1, 1980.
48. George F. Kennan, "On Nuclear War," *New York Review of Books*, XXVIII (Jan. 1981), 10; *New York Times*, Nov. 18, 1981.
49. George F. Kennan, "Cease This Madness," *The Atlantic Monthly*, CCXLVII (Jan. 1981), 25–28.
50. George F. Kennan, "Soviet-American Relations: Breaking the Spell," *The New Yorker*, LIX (Oct. 3, 1981), 44, 53.

## Chapter 8

An earlier, more extensively documented version of this chapter appeared in the *Wisconsin Magazine of History*, LVI (Winter 1972–73), 119–26, under the title "Foreign Aid Under Wraps: The Point Four Program."

1. Clark Clifford Oral History, HSTL.
2. *Public Papers, Truman, 1949*, 114–15.
3. Ben Hill Oral History, HSTL.
4. Elsey Oral History.
5. Clark Clifford to Herbert Feis, July 16, 1953, Box 36, Elsey Papers.
6. George Elsey to William Tate, probably 1952, *ibid.*
7. Memorandum by Goerge Elsey, Sept. 12, 1963, *ibid.*
8. *Public Papers, Truman, 1949*, 118–19.
9. David D. Lloyd to Clark Clifford, March 19, 1949, Box 27, Clifford Papers.
10. Walter S. Salant to Edwin Nourse, May 9, 1949, Box 2, Walter Salant Papers, HSTL.
11. Clifford Oral History.
12. *Public Papers, Truman, 1949*, 329–33.
13. Memorandum for Files, Dec. 3, 1949, Chronological File, David D. Lloyd Papers, HSTL.
14. Memorandum of Conversation, May 15, 1950, Box 68, Warren Austin Papers, Univ. of Vermont Library, Burlington, Vt.
15. Elsey notes, March 14, 1949, Box 36, Elsey Papers.
16. N. A. Bogdau in National Foreign Trade Council, *Report of 36th National Convention* (New York, 1949), 148.
17. C. V. Whitney to David D. Lloyd, June 24, 1949, Clifford Papers.
18. *Fortune*, XLI (May 1950), 26.
19. U.S. Department of State, *Point 4: What It Is . . . How It Works* (Washington, D.C., 1953), 9.
20. Vera M. Dean, *Main Trends in Postwar Foreign Policy* (New Delhi, 1950), 27.

21. Quoted in Guenther Stein, *The World the Dollar Built* (London, 1952), 252.
22. Quoted in William Burns, *Economic Aid and American Policy toward Egypt, 1955–1981* (Albany, N.Y., 1985), 46.
23. *Public Papers, Eisenhower, 1957* (Washington, D.C., 1958), 61, 63.
24. Quoted in Dennis Merrill, "Bread and the Ballot: The United States and India's Economic Development, 1947–1961" (unpublished Ph.D. diss., Univ. of Connecticut, 1986), 257.


# Chapter 9

1. Quoted in Wilbur C. Eveland, *Ropes of Sand: America's Failure in the Middle East* (New York, 1980), 101.
2. Quoted in Miles Copeland, *The Game of Nations: The Amorality of Power Politics* (New York, 1969), 146.
3. Eveland, *Ropes of Sand*, 101.
4. Quoted in Copeland, *Game of Nations*, 148.
5. Raymond Hare Oral History, Dwight D. Eisenhower Library, Abilene, Kansas (hereafter cited as DDEL).
6. Quoted in Mohamed H. Heikal, *The Cairo Documents* (Garden City, N.Y., 1973), 38.
7. Quoted in Burton I. Kaufman, "Mideast Multinational Oil, U.S. Foreign Policy, and Antitrust: The 1950s," *Journal of American History*, LXIII (March 1977), 953.
8. *Department of State Bulletin*, XXXIII (Sept. 5, 1955), 379.
9. Quoted in William R. Polk, *The United States and the Arab World*, 3rd ed. (Cambridge, Mass., 1975), 368.
10. Memorandum, "Near East Policies," John Foster Dulles to the President, March 28, 1956, Box 5, Dulles-Herter Series, Ann Whitman File, Dwight D. Eisenhower Papers, DDEL (hereafter cited as Whitman File).
11. Memorandum of conversation, Oct. 18, 1955, Box 6, Subject Series, John Foster Dulles Papers, DDEL.
12. Quoted in Heikal, *Cairo Documents*, 40.
13. Dulles quotations in U.S. Congress, Senate Committee on Foreign Relations, *Executive Sessions, Historical Series, 1953*, V (Washington, D.C., 1977), 441, 454.
14. Quoted in Gail E. Meyer, *Egypt and the United States: The Formative Years* (Rutherford, N.J., 1980), 93.
15. Quoted in *ibid.*, 120–21.
16. Quoted in Heikal, *Cairo Documents*, 51.

17. Dulles telephone call to Herbert Hoover, Jr., quoted in Carolyn A. Tyson, "Making Foreign Policy: The Eisenhower Doctrine" (unpublished Ph.D. diss., George Washington Univ., 1984), 137.

18. John C. Campbell, *Defense of the Middle East*, rev ed. (New York, 1960), 61.

19. Anthony Eden, *Full Circle* (Boston, 1960), 374–75.

20. Quoted in Mohamed Heikal, *The Sphinx and the Commissar: The Rise and Fall of Soviet Influence in the Middle East* (New York, 1978), 55.

21. Hare Oral History.

22. Quoted in Oles M. Smolansky, *The Soviet Union and the Arab East under Khrushchev* (Lewisburg, Pa., 1974), 23.

23. Telegram, John Foster Dulles to Acting Secretary, March 8, 1956, Box 5, Dulles-Herter Series, Whitman File.

24. Memorandum, "Near East Policies", March 28, 1956, Whitman File.

25. Quoted in Meyer, *Egypt and the United States*, 142.

26. Eisenhower's note to Dulles on letter of Eli Ginzberg to Eisenhower, July 17, 1956, Box 5, Dulles-Herter Series, Whitman File.

27. Quoted in Burns, *Economic Aid*, 91.

28. Telephone call to Allen Dulles, July 19, 1956, Box 5, Telephone Calls Series, Dulles Papers, DDEL.

29. Quoted in Donald Neff, *Warriors at Suez: Eisenhower Takes America into the Middle East* (New York, 1981), 262.

30. Quoted in Meyer, *Egypt and the United States*, 146.

31. Dulles to the President, Sept. 15, 1956, Box 8, Dulles-Herter Series, Whitman File.

32. Memorandum of Conversation with the President, Aug. 29, 1956, Box 5, White House Memoranda Series, Dulles Papers, DDEL.

33. Minnich notes, Legislative Leaders Meeting, Aug. 12, 1956, Box 3, Legislative Meetings Series, Office of the Staff Secretary, White House Office Records, DDEL.

34. Discussion at the 301st Meeting of the National Security Council, Oct. 26, 1956, Box 8, NSC Series, Whitman File.

35. Quoted in Chester L. Cooper, *The Lion's Last Roar: Suez, 1956* (New York, 1978), 163.

36. Telephone call, Eisenhower and Dulles, Oct. 30, 1956, Box 18, DDE Diary Series, Whitman File.

37. Memorandum of Conversation, Eisenhower, Arthur Fleming, and Andrew Goodpaster, Oct. 30, 1956, Box 19, *ibid.*

38. Quoted in Emmet John Hughes, *The Ordeal of Power* (New York, paperback ed., 1964), 192.

39. Discussion at the 302nd Meeting of the National Security Council, Nov. 1, 1956, Box 8, NSC Series, Whitman File.

40. *Ibid.*
41. Memorandum, Nov. 1, 1956, Box 19, International Series, *ibid.*
42. Quoted in Heikal, *Sphinx,* 70–71.
43. Discussion at the 303rd Meeting of the National Security Council, Nov. 8, 1956, Box 8, NSC Series, Whitman File.
44. Anwar el-Sadat, *In Search of Identity: An Autobiography* (New York, 1978), 146.
45. "Near East Policy," memorandum for the President by Herbert Hoover, Jr., Nov. 8, 1956, Box 6, Dulles-Herter Series, Whitman File.
46. Notes of Trip, Nov. 1956, Hanson Baldwin, Box 93, Baldwin Papers, Yale Univ. Library.
47. Quoted in Tyson, "Making Foreign Policy," 169.
48. Louis J. Halle to Walter Lippmann, Nov. 30, 1956, Box 75, Lippmann Papers.
49. Discussion at the 305th Meeting of the National Security Council, Nov. 30, 1956, Box 8, NSC Series, Whitman File.
50. Memorandum of Conference with the President, Nov. 21, 1956, Box 19, DDE Diary Series, *ibid.*
51. Memorandum of Conversation, Nov. 26, 1956, *ibid.*
52. John Foster Dulles in Memorandum of Conference with the President, Dec. 20, 1956, Box 20, *ibid.*
53. Discussion, 305th meeting of the National Security Council, Nov. 30, 1956, Box 8, Whitman File.
54. Cablegram, Dec. 12, 1956, Box 20, DDE Diary Series, Whitman File.
55. Telephone call from President, Dec. 8, 1956, Box 11, Telephone Calls Series, Dulles Papers, DDEL.
56. Notes on Presidential-Bipartisan Congressional Leadership Meeting, Jan. 1, 1957, Box 21, DDE Diary Series, Whitman File; Minnich notes, Box 4, Legislative Meeting Series, Office of the Staff Secretary, White House Office Records.
57. *Department of State Bulletin,* XXXVI (Jan. 28, 1957), 128.
58. Quoted in Copeland, *Game of Nations,* 216.
59. Pat M. Holt Oral History, Historical Office, U.S. Senate, Washington, D.C.
60. Senator Sam Irvin in U.S. Congress, Senate Committee on Foreign Relations, *Executive Sessions, Historical Series, 1957,* IX (Washington, D.C., 1979), 343.
61. Quoted in Cecil Crabb, *The Doctrines of American Foreign Policy* (Baton Rouge, La., 1982), 170.
62. Burns, *Economic Aid,* 110.
63. Senate, *Executive Sessions, 1957,* IX, 19.

64. Memorandum, Jan. 31, 1957, Box 41, International Series, Whitman File.
65. Memorandum, Aug. 20, 1957, Box 9, Whitman Diary Series, *ibid.*
66. Patrick Seale, *The Struggle for Syria* (London, 1965), 292.
67. Telegram 3817, Robert McClintock to Secretary of State, May 13, 1958, 783A.00/5-1358, Department of State Records, Department of State, Washington, D.C.
68. Telegram 3826, Robert McClintock to Secretary of State, May 13, 1958, *ibid.*
69. Memorandum, "Lebanese Crisis," May 13, 1958, *ibid.*
70. Telegram 4890, John Foster Dulles to American Embassy-Beirut, June 19, 1958, Box 34, International Series, Whitman File.
71. Memorandum of Conversation, "Lebanon," June 15, 1958, Box 33, DDE Diary Series, *ibid.*
72. John Foster Dulles to Henry Cabot Lodge, June 25, 1958, 783.A.00/6-2358, State Records, Department of State.
73. Assistant Secretary of State William B. Macomber in U.S. Congress, Senate Committee on Foreign Relations, *Executive Sessions, Historical Series, 1958,* X (Washington D.C., 1980), 603.
74. Memorandum of Conference with the President, July 15, 1958, Box 35, DDE Diary Series, Whitman File.
75. Staff Notes, July 15, 1958, *ibid.*
76. John Foster Dulles in Minnich notes, meeting with the President, July 18, 1958, Box 5, Cabinet Series, Office of the Staff Secretary, White House Office Records.
77. Quoted in Charles W. Thayer, "The Lebanese Landing," 3, Charles W. Thayer Papers, HSTL.
78. Quoted in Jack Shulimson, "Marines in Lebanon, 1958" (Washington, D.C., U.S. Marine Corps publication, 1983), 15.
79. Memorandum, George E. Reedy, Jr., to Senator Lyndon B. Johnson, n.d. [but July, 1958], Box 427, Office Files of George E. Reedy, Jr., Lyndon B. Johnson Senatorial Papers, Lyndon B. Johnson Library, Austin, Texas.
80. Robert Murphy, *Diplomat Among Warriors* (Garden City, N.Y., 1964), 404.

## Chapter 10

This chapter was first delivered as a lecture at the University of North Carolina, Chapel Hill, in February 1976, and later published as "Bearing the Burden: A Critical Look at JFK's Foreign Policy," *Virginia Quarterly*

*Review*, LIV (Spring 1978), 193–212. Revised and expanded through further research, it is here published with notes for the first time.

1. Quoted in Roger Hilsman, *To Move a Nation* (Garden City, N.Y., 1967), 40.
2. Walt W. Rostow, *The Diffusion of Power* (New York, 1972), 296.
3. *Public Papers, Kennedy, 1963* (Washington, D.C., 1964), 659.
4. Quoted in Stephen E. Ambrose, *Rise to Globalism*, 4th rev. ed. (New York, 1985), 201.
5. Quoted in Jim F. Heath, *Decade of Disillusionment* (Bloomington, Ind., 1975), 61.
6. Walt W. Rostow, "The Third Round," *Foreign Affairs*, XLII (Oct. 1963), 5–6.
7. CBS News, *Face the Nation, 1960–1961* (New York, 1972), VI, 408.
8. Quoted in *The New Republic*, CXLI (Nov. 2, 1959), 3.
9. Patrick Anderson, "Clark Clifford 'Sounds the Alarm,'" *New York Times Magazine*, Aug. 8, 1971, 56, 58.
10. *Public Papers, Kennedy, 1962* (Washington, D.C., 1963), 807.
11. John F. Kennedy, *The Strategy of Peace* (New York, 1960), 7.
12. Quoted in Theodore C. Sorensen, *Kennedy* (New York, 1965), 199.
13. Quoted in Heath, *Decade*, 120.
14. Quoted in Robert A. Divine, "The Education of John F. Kennedy," in Frank Merli and Theodore A. Wilson, eds., *Makers of American Diplomacy* (New York, 1974), 623.
15. David Halberstam, "The Programming of Robert McNamara," *Harper's Magazine*, CCXLII (Feb. 1971), 38.
16. Quoted in Lloyd Gardner, "Cold War Counter Revolution, 1961–1970," in William A. Williams, ed., *From Colony to Empire* (New York, 1972), 432.
17. *New York Times*, Aug. 25, 1960.
18. Quoted in Richard J. Walton, *Cold War and Counterrevolution* (Baltimore, 1973), 9.
19. Quoted in Peter Joseph, *Good Times* (New York, 1974), 4.
20. W.W. Rostow to C.D. Jackson, Feb. 17, 1961, Box 75, C.D. Jackson Papers, DDEL.
21. Quoted in Joseph, *Good Times*, 54.
22. Arthur M. Schlesinger, Jr., *A Thousand Days* (Boston, 1965), 259.
23. James Barber, *The Presidential Character* (Englewood Cliffs, N.J., 1972), 298.
24. William V. Shannon, "The Kennedy Administration," *American Scholar*, XXX (Autumn 1961), 487.
25. Memorandom for the Secretary of State, n.d. [but probably 1961],

Box 68, Departments and Agencies File, President's Office File, John F. Kennedy Papers, JFKL.

26. Quoted in Gerald T. Rice, *The Bold Experiment: JFK's Peace Corps* (Notre Dame, Ind., 1985), 59.

27. Quoted in Peter Collier and David Horowitz, *The Kennedys* (New York, 1984), 265n.

28. Quoted in Henry Fairlie, *The Kennedy Promise* (New York, 1973), 180–81.

29. Quoted in Kenneth P. O'Donnell and David F. Powers, *"Johnny, We Hardly Knew Ye"* (Boston, 1972), 287.

30. Quoted in Hugh Sidey, *John F. Kennedy, President* (New York, 1964), 127.

31. Rostow, *Diffusion*, 211.

32. Midge Decter, "Kennedyism," *Commentary*, XLIX (Jan. 1970), 21.

33. Chester Bowles Oral History, JFKL.

34. Quoted in Decter, "Kennedyism," 20.

35. Memorandum, Arthur M. Schlesinger, Jr., n.d. [but probably 1961], Box 121, President's Office File, Kennedy Papers.

36. Quoted in Charles Roberts, "Image and Reality," in Kenneth W. Thompson, ed., *The Kennedy Presidency* (Lanham, Md., 1985), 181.

37. Theodore H. White, "The Action Intellectuals," *Life*, LXII (June 1967), 43.

38. Quoted in Halberstam, "Programming," 62.

39. Quoted in Hilsman, *To Move a Nation*, 523.

40. *Public Papers, Kennedy, 1961* (Washington, D.C., 1962), 1.

41. Quoted in Heath, *Decade*, 119.

42. Quoted in John Bartlow Martin, *Adlai Stevenson and the World* (Garden City, N.Y., 1977), 634.

43. Quoted in Harris Wofford, *Of Kennedys and Kings* (New York, 1980), 426.

44. Arthur Schlesinger, Jr., "A Biographer's Perspective," in Thompson, *Kennedy Presidency*, 21.

45. *Public Papers, Kennedy, 1961*, 304–6.

46. Louis Halle to Walter Lippmann, May 3, 1961, Box 75, Lippmann Papers.

47. Quoted in Schlesinger, *A Thousand Days*, 251.

48. Quoted in Thomas Powers, *The Man Who Kept the Secrets* (New York, 1981, paperback ed.), 174.

49. Quoted in U.S. Congress, Senate Select Committee to Study Governmental Operations with Respect to Intelligence Activities, *Alleged Assassination Plots Involving Foreign Leaders: An Interim Report* (Washington, D.C., 1975), 142n.

50. Adolf A. Berle, *Navigating the Rapids, 1918–1971* (New York, 1973), 774.
51. Harold Macmillan, *At the End of the Day, 1961–1963* (New York, 1973), 183.
52. Arthur M. Schlesinger, Jr., *Robert Kennedy and His Times* (Boston, 1978), 525.
53. John Kenneth Galbraith, "The Plain Lessons of a Bad Decade," *Foreign Policy*, No. 1 (Winter 1970–71), 32.
54. Quoted in James A. Nathan, "The Missile Crisis: His Finest Hour Now," *World Politics*, XXVII (Jan. 1975), 269.
55. *Ibid.*, 269–70.
56. Rostow, *Diffusion*, 126.
57. Quoted in *ibid.*, 185.
58. Memorandum by Arthur M. Schlesinger, Jr., n.d. [but probably 1961], Box 121, Kennedy Papers.
59. "Guerrilla Warfare in Underdeveloped Areas," in Marcus G. Raskin and Barnard B. Fall, eds., *The Viet-Nam Reader*, rev. ed. (New York, 1967), 113.
60. Quoted in Seyom Brown, *The Faces of Power*, 2nd ed. (New York, 1983), 153.
61. Memorandum for the President by Walt W. Rostow, March 29, 1961, Box 193, National Security Files, Kennedy Papers.
62. Quoted in Bruce Miroff, *Pragmatic Illusions* (New York, 1976), 146.
63. Arthur M. Schlesinger, Jr., "JFK Plus 20: What the Thousand Days Wrought," *The New Republic*, CLXXXIX (Nov. 21, 1983), 25.
64. Quoted in Louise FitzSimons, *The Kennedy Doctrine* (New York, 1972), 15.
65. Anthony Hartley, "John Kennedy's Foreign Policy," *Foreign Policy*, No. 4 (Fall 1971), 86.
66. John Kenneth Galbraith, *Ambassador's Journal* (Boston, 1969), 311.

## Chapter 11

This chapter is based upon "Isolationism Revisited," first published in *The Nation*, CCIX (Sept. 1, 1969), 166–69. Senator J. William Fulbright had the article reprinted in the *Congressional Record*, CXV (Sept. 12, 1969), S10506–8.

1. Robert Komer in W. Scott Thompson and Donaldson D. Frizzell, eds., *The Lessons of Vietnam* (New York, 1977), 211.
2. Quoted in Ronald Steel, *Walter Lippmann and the American Century* (Boston, 1980), 586.

3. *Public Papers, Nixon, 1969* (Washington, D.C., 1971), 433.
4. Selig Adler, *The Isolationist Impulse* (New York, 1957, paperback ed.), 242.
5. Manfred Jonas, *Isolationism in America, 1935–1941* (Ithaca, N.Y., 1966), 3, viii.
6. *Congressional Record*, LXXXI (March 3, 1937), 1785.
7. Louis Ludlow, *Hell or Heaven* (Boston, 1937), 142–43, 147.
8. Norman Thomas and Bertram D. Wolfe, *Keep America Out of War* (New York, 1939), 20.
9. *Ibid.*, 50.
10. Quoted in Wayne S. Cole, *Senator Gerald P. Nye and American Foreign Relations* (Minneapolis, 1962), 158.
11. Statement by Dwight MacDonald, John Chamberlain, Sidney Hook, and others, quoted in Michael Wreszin, *Oswald Garrison Villard* (Bloomington, Ind. 1965), 257.
12. Quoted in Warren I. Cohen, *The American Revisionists* (Chicago, 1967), 192–3.
13. Quoted in Jonas, *Isolationism*, 268.
14. Quoted in Cole, *Nye*, 191.
15. Charles A. Beard, "Giddy Minds and Foreign Quarrels," *Harper's Magazine*, CLXXIX (Sept. 1939), 341.
16. Quoted in Cole, *Nye*, 63.
17. John Wiltz, "The Nye Committee Revisited," *The Historian*, XXIII (Feb. 1961), 231–2.
18. Quoted in Paul Holbo, ed., *Isolationism and Interventionism, 1932–1941* (Chicago, 1967), 42–43.
19. Quoted in Robert A. Divine, *The Illusion of Neutrality* (Chicago, 1962), 318; Cole, *Nye*, 126–27.
20. Quoted in Gabriel Kolko, "American Business and Germany, 1930–1941," *Western Political Quarterly*, XV (Sept. 1962), 726.
21. Jonas, *Isolationism*, 114.
22. Charles Lindbergh, *Autobiography of Values* (New York, 1978), 194.

## Chapter 12

This topic was first explored in "After Peking, Moscow: New Levers of Containment", *The Nation*, CCXIV (April 24, 1972) and was developed further in "The Perils of a Grand Design, 1969–1977," Chapter 15, Volume 2 of *American Foreign Policy: A History*, 3rd ed. (Lexington, Mass.: D. C. Heath and Company, 1987). The latter publication was co-authored with J. Garry Clifford and Kenneth J. Hagan.

1. Richard M. Nixon, *RN: The Memoirs of Richard Nixon* (New York, 1978), 611.
2. Quoted in William Safire, *Before the Fall: An Inside View of the Pre-Watergate White House* (Garden City, N.Y., 1975), 102.
3. Nixon, *RN*, 529.
4. H. R. Haldeman, *The Ends of Power* (New York, 1978), 98.
5. Henry A. Kissinger, *White House Years* (Boston, 1979), 55.
6. Harry Schwartz in *New York Times*, Feb. 21, 1972.
7. *Department of State Bulletin*, LXVI (March 20, 1972), 421.
8. *Ibid.*, 437.
9. Michael Mandelbaum, *The Nuclear Question* (Cambridge, England, 1979), 218.
10. *Public Papers, Richard M. Nixon, 1969* (Washington, D.C., 1971), 19.
11. Kissinger, *White House Years*, 1245.
12. Quoted in Richard J. Barnet, *The Giants* (New York, 1977), 102.
13. Quoted in Robert W. Stookey, *America and the Arab States* (New York, 1975), 221.
14. Quoted in William B. Quandt, *Decade of Decisions: American Policy Toward the Arab-Israeli Conflict, 1967–1976* (Berkeley, 1977), 127.
15. Nixon, *RN*, 924.
16. Quoted in Quandt, *Decade of Decisions*, 188.
17. Quoted in Edward R. F. Sheehan, "How Kissinger Did It: Step by Step in the Middle East," *Foreign Policy*, No. 22 (Spring 1976), 48.
18. Quoted in *New York Times*, March 15, 1976.
19. Kissinger, *White House Years*, 1264.
20. *Department of State Bulletin*, LXXII (June 2, 1975), 713.
21. Quoted in Elain P. Adam and Richard P. Stebbins, eds., *American Foreign Relations, 1976* (New York, 1978), 13.

## Chapter 13

This chapter was originally published with fuller documentation in Michael Barnhart, ed., *Congress and United States Foreign Policy: Controlling the Use of Force in the Nuclear Age* (Albany, N.Y.: State University of New York Press, 1987), 154–175. The following, although not cited in the notes, were particularly helpful in studying the organization of the intelligence community and in tracing the legislative history of oversight: Mark M. Lowenthal, *U.S. Intelligence: Evolution and Anatomy* (New York, 1984) and several "Issue Briefs" prepared by Lowenthal and others for the Library of Congress, Congressional Research Service.
1. Quoted in *Time*, CXIII (April 23, 1984), 16.

2. Ohman cartoon, *The Portland Oregonian* in *Washington Post National Weekly Edition*, Jan. 14, 1985.

3. U.S. Congress, House Permanent Select Committee on Intelligence, *Congressional Oversight of Covert Activities* (Hearings, Washington, 1984), 24.

4. Quoted in John Stockwell, *In Search of Enemies: A CIA Story* (New York, 1978), 235.

5. Stanfield Turner, *Secrecy and Democracy: The CIA in Transition* (Boston, 1985), 188.

6. Quoted in Arthur M. Schlesinger, Jr., *The Imperial Presidency* (Boston, 1973), 318.

7. Quoted in U.S. Congress, Senate Committee on Government Operations, *Oversight of U.S. Government Intelligence Functions* (Hearings, Washington, D.C., 1976), 81.

8. Quoted in David Wise, "Covert Operations Abroad: An Overview," in Robert L. Borosage and John Marks, eds., *The CIA File* (New York, 1976), 18.

9. Quoted in Harry H. Ransom, *The Intelligence Establishment* (Cambridge, Mass., 1970), 169.

10. Senate, *Oversight*, 334.

11. Representative Robert N. Giaimo et al. to colleagues, July 9, 1975, Box 125, Series IV, Robert N. Giaimo Papers, Univ. of Connecticut Library, Storrs, Conn.

12. House, *Congressional Oversight*, 30.

13. John Maury, "CIA and the Congress," n.d. [but probably 1973 or 1974], Files of Central Intelligence Agency, Washington, D.C.

14. William Colby and Peter Forbath, *Honorable Men: My Life in the CIA* (New York, 1978), 401.

15. Quoted in Ransom, *Intelligence Establishment*, 167.

16. Quoted in John V. Lindsay, "An Inquiry into the Darkness of the Cloak, the Sharpness of the Dagger," *Esquire*, LXI (March 1964), 109.

17. Maury, "CIA and the Congress," 12.

18. Quoted in Norman D. Sandler, *28 Years of Looking the Other Way: Congressional Oversight of the Central Intelligence Agency, 1947–1975* (Cambridge, Mass., 1975), 118.

19. Quoted in *ibid.*, 184–5.

20. Michael J. Harrington to Carl Albert, July 8, 1975, Box 126, Series IV, Giaimo Papers.

21. Ralph W. McGehee, *Deadly Deceits: My 25 Years in the CIA* (New York, 1983), 82–84.

22. Quoted in Loch K. Johnson, *A Season of Inquiry: The Senate Intelligence Investigation* (Lexington, Ky., 1985), 6.

23. Quoted in Sandler, *28 Years*, 94–95.

24. Quoted in Fred Branfman, "The President's Secret Army: A Case Study—the CIA in Laos, 1962–1972," in Borosage and Marks, *CIA File*, 73.

25. Quoted in U.S. Congress, Senate Select Committee to Study Governmental Operations with Respect to Intelligence Activities, *Final Report: Book I* (Washington, D.C., 1976), 367n.

26. Quoted in Louis Fisher, *Presidential Spending Power* (Princeton, N.J., 1975), 218.

27. Quoted in Gary Sperling, "Central Intelligence and Its Control: Curbing Secret Power in a Democratic Society," in *Congressional Record*, CXII (July 14, 1966), 15764.

28. Senator Carl Hayden quoted in Ransom, *Intelligence Establishment*, 166.

29. Senate, *Oversight*, 25.

30. Memorandum of Conversation, Bipartisan Leaders Breakfast with the President, May 26, 1960, Box 25, Office of the Staff Secretary, White House Office Records, DDEL.

31. Quoted in Ransom, *Intelligence Establishment*, 172.

32. *Congressional Record*, CXII (July 14, 1966), 15677.

33. Michael J. Harrington et al. *The CIA: Past Transgressions and Future Controls: A Symposium* (Providence, R.I., 1975), 5.

34. Quoted in Dean D. Welch, "Secrecy, Democracy and Responsibility: The Central Intelligence Agency and Congress" (unpublished Ph.D. diss., Vanderbilt Univ., 1976), 193.

35. *Congressional Record*, CXX (Oct. 2, 1974), 33479.

36. Senate, *Final Report: Book I*, 151.

37. Thomas K. Latimer, "U.S. Intelligence and the Congress," *Strategic Review*, VII (Summer 1979), 49.

38. Turner, *Secrecy*, 87.

39. U.S. Congress, Senate Select Committee on Intelligence, *National Intelligence Act of 1980* (Hearings, Washington, D.C., 1980), 170.

40. Carter, *Keeping Faith*, 511, 518.

41. U.S. Congress, Senate Select Committee on Intelligence, *Intelligence Oversight Act of 1980* (Washington, D.C., 1980).

42. John M. Oseth, *Regulating U.S. Intelligence Operations* (Lexington, Ky., 1985), 170.

43. Norman Y. Mineta quoted in Johnson, *A Season*, 263.

44. Unidentified. Quoted in Jay Peterzell, "Can Congress Really Check the CIA?" *Washington Post*, April 24, 1983, C4.

45. *Ibid.*

46. Unidentified. Quoted in *Newsweek*, CIII (April 23, 1984), 24.

47. *Ibid.*

48. U.S. Congress, Senate Select Committee on Intelligence, *Report* No. 98-665 (Washington, D.C., 1985), 5.
49. Quoted in *New York Times*, Aug. 9, 1985.
50. Quoted in *Hartford Courant*, Aug. 28, 1985.
51. Comment by Stanley Heginbotham, Jacob K. Javits Collection Inaugural Conference on Congress and United States Foreign Policy, State Univ. of New York at Stony Brook, Oct. 25, 1985.
52. Quoted in *New York Times*, Oct. 18, 1983.
53. Quoted in Johnson, *A Season*, 8.
54. Comment by Senator John Kerry at Javits Conference, Oct. 26, 1985.
55. Quoted in *New York Times*, July 7, 1986.

## Chapter 14

This chapter was delivered as the Presidential Address to the Society for Historians of American Foreign Relations in December 1987 in Washington, D.C., and, in different form, was published as "Historical Memory and Illusive Victories: Vietnam and Central America", *Diplomatic History*, XII (Winter 1988).

1. Quoted in Strobe Talbott, *The Russians and Reagan* (New York, 1984), 32; *Public Papers, Reagan, 1981* (Washington, D.C., 1982), 57.
2. Quoted in Hedrick Smith, "Reagan: What Kind of World Leader? *New York Times Magazine*, Nov. 16, 1980, 172.
3. Quoted in Harold Molineu, *U.S. Policy Toward Latin America* (Boulder, Colo., 1986), 176.
4. Quoted in *Department of State Bulletin*, LXXXI (March 1981), 7; *New York Times*, March 19, 1981.
5. Quoted in George C. Herring, "Vietnam, El Salvador, and the Uses of History," in Kenneth M. Coleman and George C. Herring, eds., *The Central American Crisis* (Wilmington, Del., 1985), 99.
6. Quoted in David Fromkin and James Chace, "What *Are* the Lessons of Vietnam?" *Foreign Affairs*, LXIII (Spring 1985), 724.
7. Quoted in *New York Times*, Aug. 19, 1980.
8. Quoted in John C. Pratt, ed., *Vietnam Voices* (New York, 1984), 668.
9. Quoted in Gaddis, *Strategies*, 345.
10. William C. Westmorland, "Vietnam in Perspective," *Military Review*, LIX (Jan. 1979), 42.
11. Quoted in *Washington Post National Weekly Edition*, April 29, 1985.
12. Quoted in George C. Herring, "The 'Vietnam Syndrome' and American Foreign Policy," *Virginia Quarterly Review*, LVII (Fall 1981), 595.

13. Nixon, *RN*, 347.
14. Roger Hilsman, "Must We Invade the North?" *Foreign Affairs*, XLVI (April 1968), 430.
15. Quoted in Doyle McManus, *Free at Last!* (New York, 1981), 16.
16. Quoted in George McT. Kahin, *Intervention* (New York, 1986), 372.
17. Quoted in "After Vietnam: This War Was Different," *Washington Post*, May 25, 1980.
18. Quoted in Earl C. Ravenal, *Never Again* (Philadelphia, 1978), 25.
19. James C. Thomson, Jr., "Minutes of a White House Meeting, Summer, 1967," in Robert Manning and Michael Janeway, eds., *Who We Are* (Boston, 1975), 43–44.
20. Quoted in James Fallows, *National Defense* (New York, 1981), 26.
21. Quoted in *Washington Post*, July 6, 1981.
22. Philip A. Caputo, *A Rumor of War* (New York, 1977; paperback ed.), 100.
23. *Ibid.*, 69.
24. Bernard W. Rogers, U.S. Department of the Army Vietnam Studies, *Cedar Falls-Junction City* (Washington, D.C., 1974), 41.
25. Quoted in George McT. Kahin and John W. Lewis, *The United States in Vietnam*, rev. ed. (New York, 1969), 373.
26. Quoted in Leslie H. Gelb and Richard K. Betts, *The Irony of Vietnam* (Washington, D.C., 1979), 169–70.
27. Quoted in George C. Herring, *America's Longest War*, 2nd ed. (New York, 1986), 155.
28. John E. Mueller, "The Search for the 'Breaking Point' in Vietnam," *International Studies Quarterly*, XXIV (Dec. 1980), 497–519.
29. Quoted in William Broyles, Jr., *Brothers in Arms* (New York, 1986), 267.
30. Bruce Palmer, Jr., *The 25-Year War* (Lexington, Ky., 1984), 176.
31. Quoted in Michael Maclear, *The Ten Thousand Day War* (New York, 1981), 354.
32. Quoted in William M. LeoGrande, "A Splendid Little War: Drawing the Line in El Salvador," *International Security*, VI (Summer 1981), 45.
33. Quoted in *New York Times*, July 2, 1981.
34. Quoted in Sidney Blumenthal, "Marketing the President," *New York Times Magazine*, Sept. 13, 1981, 112.
35. Quoted in Marvin E. Gettleman et al. eds., *El Salvador* (New York, 1981), 355.
36. Quoted in *New York Times*, May 28, 1981.
37. Quoted in *ibid.*, Aug. 13, 1986.

# Index

303